T0300710

# PRAISE FOR *IS EARTH EXCEPTIONAL?*

"Livio and Szostak adeptly describe how astrophysics, geology, and chemistry may have collided on the primordial Earth to spark life. Provocative and thoughtful, they tackle the age-old question: Is there life in the Universe beyond Earth? You'll need to read the book to find the answer."

—Thomas R. Cech, Nobel laureate and author of *The Catalyst*

"*Is Earth Exceptional?* connects current research about two questions, 'Is life out there?' and 'How did life start here?' to where they meet in the middle in a way that is exciting and accessible. This is the biggest human quest: to understand our place in the Universe. I highly recommend this book. . . . It is exceptional."

—Adam Riess, Nobel laureate and Bloomberg Distinguished Professor, Johns Hopkins

"*Is Earth Exceptional?* is a mesmerizing exploration of life's origins and its potential beyond our planet. With profound insights from astrophysicist Livio and my PhD advisor and Nobel laureate Szostak, this is a must-read for anyone curious about the universe and our place within it."

—Jennifer Doudna, Nobel laureate and coauthor of *A Crack in Creation*

"This book offers the clearest account I have read anywhere on our current understanding of how life could have begun from simpler chemicals, and it also explores the prospects for extraterrestrial life. These are two of the biggest unsolved mysteries in science, and the book makes for fascinating reading."

—Venki Ramakrishnan, Nobel laureate and former president of the Royal Society

"How did life begin? Are we alone in the cosmos? These are eternally fascinating mysteries. But what's so exciting today is that novel insights and improved instruments are generating real scientific progress toward solving them. General readers should be grateful to the two authors of this book—distinguished in astronomy and in biochemistry—for expounding current debates and future prospects with authority and clarity. Their timely book shows how science fiction is turning into real science: it deserves wide readership."

—Martin Rees, UK astronomer royal and bestselling author of *If Science Is to Save Us*

"At long last, a popular science book on the origins of life that delves deep into the underlying science—take the dive and be richly rewarded!"

—John D. Sutherland, FRS, Darwin medalist,
MRC Laboratory of Molecular Biology, Cambridge, UK

"A masterful and insightful traverse from historical milestones to the forefront of modern science, *Is Earth Exceptional?* offers an indispensable guide for anyone seeking to understand the puzzles and mysteries surrounding the beginnings of life and its potential beyond Earth."

—Sara Seager, professor of physics, planetary science,
aeronautics, and astronautics, MIT

"How could the geochemistry of young Earth lead to biochemistry? Is life a common occurrence in the cosmos? In this brilliant new book, Livio and Szostak bring us to the threshold of an imminent breakthrough. They are by far the best guides to take us on this quest. Eloquent, deep, and refreshingly up to date, the book is a fascinating journey through the breathtaking progress of the past decade. It is *The Da Vinci Code* of science books—a remarkable page-turner to the very end."

—Dimitar Sasselov, Phillips Professor of Astronomy, Harvard University, and
director of the Harvard Origins of Life Initiative

"Livio and Szostak have written a comprehensive and compelling review of the evidence of how the origin of life might have happened and the chances of finding life elsewhere in the Universe. It's a great combination of the chemistry and biology of life's origins by one of the top researchers in the field and a bestselling astrophysicist. They guide us through the exciting breakthroughs in exobiology and the potential for extraterrestrial life. A fascinating read to a general reader looking for answers to the question: Are we alone in the Universe? Enjoyable and highly informative."

—J. Craig Venter, leader of the teams that sequenced the first
human genome and built the first synthetic cell

"This is an exceptional book on an exceptional question. With astronomers finally poised to begin the search for alien life in earnest, Livio and Szostak offer a beautifully written and accessible tour of the essential questions surrounding that quest. From how life begins to which planets it might begin on, *Is Earth Exceptional?* delivers deep insights into the cutting-edge science needed to answer those questions. This is a must-read book for anyone interested in the ancient question of life and its cosmic fecundity."

—Adam Frank, professor of astrophysics and author of
*The Little Book of Aliens*

"A mind-blowing ramble through the RNA world and beyond that confronts the ultimate challenge for the origins of life: Are we alone in the Universe?"

—Joseph Silk, author of *Back to the Moon*

# IS EARTH EXCEPTIONAL?

## THE QUEST FOR COSMIC LIFE

MARIO LIVIO & JACK SZOSTAK

BASIC BOOKS
NEW YORK

Basic Books
Hachette Book Group
1290 Avenue of the Americas, New York, NY 10104
www.basicbooks.com

Printed in Canada

First Edition: September 2024

Published by Basic Books, an imprint of Hachette Book Group, Inc. The Basic Books name and logo is a trademark of the Hachette Book Group.



Print book interior design by Sheryl Kober.

Library of Congress Cataloging-in-Publication Data

Names: Livio, Mario, 1945– author. | Szostak, Jack W., author.
Title: Is Earth exceptional? : the quest for cosmic life / Mario Livio and Jack Szostak.
Description: First edition. | New York : Basic Books, 2024. | Includes bibliographical references and index.
Identifiers: LCCN 2024002954 | ISBN 9781541602960 (hardcover) | ISBN 9781541602977 (ebook)
Subjects: LCSH: Exobiology. | Life on other planets. | Life—Origin. | Evolution (Biology) | Earth (Planet)
Classification: LCC QH326 .L58 2024 | DDC 576.8/3—dc23/eng/20240408
LC record available at https://lccn.loc.gov/2024002954

ISBNs: 9781541602960 (hardcover), 9781541602977 (ebook)

MRQ-T

10 9 8 7 6 5 4 3 2 1

# CONTENTS

# CHAPTER 1

# A Freak Chemical Accident or a Cosmic Imperative?

*It's as large as life, and twice as natural!*
—Lewis Carroll, *Through the Looking Glass*

In our everyday lives, we are used to the fact that the direction of our psychological "arrow of time" allows us to examine, study, reflect upon, and remember events in the past. We are equally aware of the fact that we cannot remember the future. We can at best attempt to make predictions, speculate about it, or envisage the future in the eye of our imagination. As poet Kahlil Gibran so expressively wrote, "For life goes not backward nor tarries with yesterday."

Somewhat paradoxically, when it comes to the phenomenon of biological life on Earth, we are quite certain about how Mother Nature will ultimately end it, in the distant future, but we don't know how exactly it started. The natural (not caused by self-destructive actions of our currently dominant species) termination of life as we know it will be dictated by relatively well-understood and predictable astrophysical and atmospheric processes (unless unforeseeable cosmic events, such as an asteroid impact or a nearby gamma-ray burst, act to bring about a premature end).

We know, for instance, that in approximately five billion years, as our Sun expands tremendously to become a red giant star, Earth will be scorched, and may even be engulfed by the Sun's expanding envelope. Complex multicellular life will become extinct much earlier, about a billion years hence, as Earth's biosphere perilously declines due to the rising temperatures associated with the late stages in the Sun's evolution.

The *origin* of life, on the other hand, is still veiled in mystery. While enormous progress has been achieved in understanding the building blocks of biology, we still don't know what precisely it was that caused life to spontaneously emerge, or how the very first cells suddenly came into existence. As British chemist John Sutherland puts it, all we can say about that momentous point in time at which chemistry gave birth to biology is that life appeared "out of the blue." Wittily, Sutherland was also referring to the "blue" associated with cyanide, which, as we shall see, played a crucial role in the origin of life.

Intimately related to the origin-of-life puzzle is another question, which has intrigued humans at least since the time of the Pythagoreans of ancient Greece: Are we alone in the universe? Or, in its more modern, somewhat more practical incarnation: Is our galaxy as filled with life as many sci-fi creations would have you believe? In other words, we would like to know whether humanity is finally about to end the loneliness of its sojourn in the Milky Way.

While one of us is an astrophysicist and the other a chemist-biologist, we have both been fascinated by these cosmic riddles throughout our entire scientific careers. We have been intrigued by these questions, yes, but for quite some time we couldn't do much more than wonder, because until fairly recently these questions were deemed to be intractable, insoluble in our lifetime, perhaps even only on the fringe of science. They tended to be relegated to that "far too difficult" category.

This situation has dramatically changed in the last three decades. The attempts to answer these precise questions—How did life on Earth begin? Are we alone in the Milky Way?—have become two of the most vibrant and dynamic frontiers of scientific investigation.

Remarkably, the answers to these inquiries hinge on a third question, one that is relatively simple to formulate, certainly well defined, and most definitely answerable (in principle at least): *How likely is it for life to emerge on the surface of a potentially habitable planet?*

This last question is being addressed by two completely distinct and largely independent lines of research. First, current laboratory studies are aimed at determining whether biology can indeed emerge from pure chemistry. Second, much of astronomy is devoted to the search for unambiguous signs of life on other planets or moons (either in the solar system or around other stars). Both of these approaches are currently attracting strong interest and are the subjects of enthusiastic efforts by dedicated communities of scientists. In fact, the search for life on planets around stars other than the Sun—extrasolar planets—is now a consensus goal of the astronomical community in the United States, as outlined in a report released in November 2021 by the National Academies of Sciences, Engineering, and Medicine. We, the authors, humbly participate (each one in his own discipline) in these quests.

One of the key points we want to highlight in this book is that pursuing the origin of life on Earth and the search for extraterrestrial life are two scientific aspirations with a powerful, symbiotic relationship. Success in one would provide an extremely encouraging clue and a strong motivation for the other. The reason is simple. If we can find a pathway to life from chemistry in the lab, it means that there is a fair chance that Nature, with its huge arsenal of diverse environments and eons at its disposal, can perhaps do it too, maybe even at numerous places in the cosmos, including in our home galaxy, the Milky Way. Moreover, if we could comprehensively understand a compelling sequence of events, processes, and environmental conditions that may have been involved in the origin of life on Earth, we could much better gauge how likely or unlikely it is for life to spontaneously pop up on other planets or moons. Such insights could therefore guide our search for alien life.

Thinking in the other direction, if we were to discover via astronomical observations that extraterrestrial life is relatively common, this would significantly reinforce our conviction that an inevitable geochemical path to life exists. This confidence, in turn, would strongly motivate the efforts to discover the right initial conditions, seed materials, necessary energy sources, and network of chemical reactions that could serve as prerequisites for life to emerge. Even more broadly, thorough examinations of the problems involved in the origin of life on the one hand, and the search for extraterrestrial life on the other, offer a unique opportunity to explore a wide range of fields and disciplines, from astronomy and geology to chemistry and biology.

There is another important point to consider here. We know that in many domains and circumstances the so-called zero-one-infinity principle (ZOI) applies. That is, an entity should either be entirely forbidden; so rare that only one specimen should be permitted; or a very large number of exemplars should be expected. If some form of alien life, totally independent of life on Earth, were to be readily discovered (what has been dubbed a *second genesis*), this would imply (applying the ZOI) that it is reasonable to assume that there are virtually infinite examples of life in the universe.

This book tells the story of those two fascinating, parallel endeavors: one with the explicit goal of finding a path from chemistry to life in the lab, the other aiming to discover extraterrestrial life. These quests are implicitly cooperating, occasionally competing (for who will get to their objective first), but always engrossing, and they complement each other in their eagerness to solve puzzles that are central to us being human—Where did we come from? Why are we here? Are we alone? In other words, at the risk of sounding a bit bombastic, the ultimate desire of these pursuits is literally to understand our origins and our place in this vast, old, and intricate cosmos.

## Life—What a Concept

While the questions of "How did life begin?" and "Is there extraterrestrial life?" have enthralled humans since ancient times, throughout most of recorded history almost everyone believed that the answer to the first question was simple: "God created it." In fact, all the way up to the beginning of the nineteenth century, even scientists were quite convinced that living things had to be endowed with some quasi-mystical "vitalism" that set them apart from inanimate matter. The second question, on the other hand, generated a debate, with wild speculations dating back millennia, either for or against the idea of a "plurality of inhabited worlds." For example, already in the first century BCE, the Roman Epicurean poet Titus Lucretius Carus wrote,

> *Why then confess you must*
> *That other worlds exist in other regions of the sky,*
> *And different tribes of men, kinds of wild beasts.*

An obvious milestone in this theoretical wrangle was Copernicus's heliocentric model, since it provided not only an entirely new perspective on Earth's significance in the grand cosmic scheme of things, but also a realistic framework within which the existence of other Earth-like worlds became at the very least imaginable. Expanding upon those, then novel, Copernican concepts, Italian Dominican friar and philosopher Giordano Bruno famously conjectured at the end of the sixteenth century that "in space, there are countless constellations, suns and planets; we see only the suns because they give light; the planets remain invisible, for they are small and dark. There are also numberless earths circling around their suns, no worse and no less than this globe of ours." Bruno's insightful imagination anticipated modern science and took him even further, to conclude: "No reasonable mind can assume that heavenly bodies that may be far more magnificent than ours would not bear upon them creatures similar or even superior to those upon our human earth." Tragically, as a result of the tenacity with which he defended other unorthodox ethical and theological ideas, then considered heretical, Bruno was burned at the stake by the Roman Inquisition on February 17, 1600.

In the seventeenth century, others started to make claims related to cosmic pluralism. Prominent scientists, such as astronomers Johannes Kepler and Christiaan Huygens, and other influential intellectuals, such as French science writer Bernard Le Bovier de Fontenelle, did not hesitate to advocate for the existence of extraterrestrial beings. After Galileo Galilei discovered four moons orbiting Jupiter, Kepler was quick to deduce: "The conclusion is quite clear. Our Moon exists for us on the Earth, not for the other globes. Those four little moons exist for Jupiter, not for us. Each planet in turn, together with its occupants, is served by its own satellites. From this line of reasoning, we deduce with the highest degree of probability that Jupiter is inhabited." Galileo himself, on the other hand, was rather agnostic about the plurality of inhabited worlds, cautiously remarking: "I should for my part neither affirm it [life on other planets] nor deny it, but should leave the decision to wiser men than I."

Concomitantly with the voices adopting the position of cosmic pluralism, there were equally vociferous denials of the claims about the existence of extraterrestrial life. The opposing views arose primarily on account of the fact that the mere idea of there being inhabitants of other planets was pregnant with potentially disturbing implications for certain doctrines of the Catholic Church. The naysayers raised such ecclesiastical conundrums as "If there were indeed people on other worlds, had they descended from Adam and Eve too?" Or, "Was Jesus Christ their Savior as well?"

Given the great influence of religious ideas throughout much of human history, it should come as no surprise that both the belief in "vitalism" and the notion that life must pervade the cosmos were initially based more on theological rather than scientific arguments. Vitalism was largely inspired by a literal interpretation of the biblical text: "And the Lord God formed man of the dust of the ground, and breathed into his nostrils the *breath of life* [emphasis added]; and man became a living soul." Aristotle also insisted that a soul is "the actuality of a body that has life." Still on the basis of religious beliefs, some nineteenth-century thinkers held out for the existence of extraterrestrial inhabited worlds, because otherwise the enormous vastness of space seemed to be a huge waste of the Creator's endeavors.

In the twentieth century, philosophers, and scientists when they waxed philosophical, embarked on numerous attempts to *define* life. Even Erwin Schrödinger, one of the founders of quantum mechanics, published in 1944 a small book entitled *What Is Life?*, which inspired enthusiasm for discovering the chemical basis of heredity. Overall, however, the endeavors to define life resulted in almost as many definitions as there were definers. Molecular biophysicist Edward Trifonov collected 123 definitions by many researchers, and after analyzing their vocabulary came up in 2011 with what he regarded as the consensus distilled definition: "Life is self-reproduction with variations." An earlier definition, which like most others generated considerable debate, was nevertheless adopted by NASA's astrobiology division: "Life is a self-sustaining chemical system capable of Darwinian evolution." What we are interested in here, however, is not a universal definition of life. We feel that on the whole, the "What is life?" discussion has not been particularly fruitful in helping us to understand the origin of life. It has also been mired in the confusion that arises from using one word to encompass multiple distinct phenomena. Rather, what we think truly matters is identifying a pathway via which biology can emerge out of the conditions on a young planet. The challenge of uncovering this elusive route is amplified by the fact that to date we only know of one example of life in the entire universe—life on Earth. Life elsewhere may, in principle, take forms that we wouldn't recognize or maybe even can't conceive of.

To make progress, biologists have identified a few essential elements that appear to be required for life, and a small number of attributes that characterize (and are crucial to) at least every life-form on Earth. The required ingredients are (1) an *energy* source to power metabolic reactions, (2) a *liquid solvent* that can facilitate such (and other) reactions, and (3) *nutrients* needed to create biomass.

The properties that characterize life on Earth are as follows: (i) life is composed of *cells*, (ii) it can perform *metabolism* (that is, it can harvest energy and materials from its environment and use them for growth and reproduction), (iii) it utilizes *catalysts* to assist and speed up chemical reactions, and (iv) it contains an *informational* system. The last property means that life can reproduce its own characteristics, and that it can undergo Darwinian evolution—it has the chemical instructions for operations, and information that can be passed on from one generation to the next. In short, life *as we know it* needs to somehow seamlessly integrate the four subsystems of compartmentalization (cells), metabolism, catalysis, and genetics.

While all origin-of-life researchers agree that these features are shared by all living things on Earth, for several decades those same researchers disagreed, and even vigorously argued, about the question of whether one of these properties is the most fundamental, and if so, which one. Specifically, which characteristic had to appear on Earth first, to allow for life to emerge? As we shall soon see, this particular muddle appears to have been resolved during the past two decades, in a somewhat unexpected way.

## The Book of Life

In Oscar Wilde's play *A Woman of No Importance*, Lord Illingworth declares: "The Book of Life begins with a man and a woman in a garden." To which Mrs. Allonby wittily responds: "It ends with Revelations."

In spite of the strong religious and emotional attachment to the notion that life had to contain some extra magic or divine intervention, opinions started to change at the beginning of the nineteenth century. A step toward freeing life from the need for a "vital force" beyond the understanding of science was taken in 1828, when German chemist Friedrich Wöhler accidentally succeeded in synthesizing urea—a substance found in urine that had previously been thought to be unique to life—from common chemicals. Delighted with his success in imitating nature in the laboratory, the ecstatic Wöhler wrote to his teacher and collaborator, chemist Jöns Jacob Berzelius: "I can no longer, so to speak, hold my chemical water and must tell you that I can make urea without needing a kidney, whether of man or dog; the ammonium salt of cyanic acid is urea."

The correspondingly dramatic leap in the understanding of biology came with Charles Darwin's theory of evolution by means of natural selection. Whereas Darwin's theory itself ducked the origin-of-life question altogether, saying absolutely nothing about how the first organisms came into being, in 1871 Darwin mused in a letter to his friend Joseph Dalton Hooker about how life on Earth might have started. He famously wrote: "If (and oh, what a big if) we could conceive in some warm little pond with all sort of ammonia and phosphoric salts, light, heat, electricity, etc. present, that a protein compound was chemically formed ready to undergo still more complex changes, at the present day such matter would be instantly devoured or absorbed, which would not have been the case before living creatures were formed!"

Darwin's prescient speculation is remarkable for no fewer than five reasons. First, it totally disposes of the need for anything supernatural in the origin of life. Second, it suggests that life may have originated in a "warm little pond," a view that, as we shall see, is stunningly compatible with our thinking today. Third, it identifies ammonia and phosphates (compounds containing nitrogen and phosphorus) as being (potentially) necessary materials for life, again an incredible foresight. Fourth, it proposes that some form of "protein compound" may have played a role in the chemistry leading to life. And fifth, to avoid the impression that living organisms may be repeatedly springing into existence, Darwin points out that the conditions under which the first life-forms emerged no longer exist today.

This idea—that life is nothing more than a combination of highly sophisticated chemical systems—is one that was initially abhorrent to quite a few people. Life, those skeptics proclaimed, is far too cleverly contrived simply to have arisen through processes of chance, while obeying only the laws of physics and chemistry. Consequently, even many of those who were willing, in principle, to accept a chemical origin of life used still to think that some incredibly rare chance event must have been required, to bring together in one fell swoop all the components of the first living cells.

The view of creating complexity all at once from a chaotic soup of simple building blocks was further motivated by the mind-boggling intricacy of all cellular life on Earth today. The most puzzling aspect of this convolution is that all the parts and processes of extant life depend on all the other parts and processes in a circular way. For example, a complex metabolism is needed to make the biochemicals that are required for the assembly of those protein enzymes that are needed to catalyze the reactions of . . . metabolism itself! Similarly, the nucleic acid molecules, DNA and RNA, are needed to encode the information that specifies the assembly of proteins—the workhorse molecules of life—which are required to make . . . yes, you guessed it, DNA and RNA. To make matters even more perplexing, to allow all of these molecules to accomplish their tasks, they need cell membranes that keep all the molecular players boxed together. But cell membranes are made of fatty compounds known as lipids, and those are synthesized by protein enzymes. This type of self-referential or recursive activity (reminiscent of a famous drawing by graphic artist M. C. Escher in which two hands are drawing each other) is so deeply embedded in the fabric of modern living organisms, that for many years it seemed that some miraculous event would have been required to bridge the gap between a random mixture of chemicals and the highly organized structure of a living cell. Even as late as 1981, Francis Crick, the co-discoverer of the double-helix structure of DNA, emphasized that "an honest man, armed with all the knowledge available to us now, could only state that in some sense, the origin of life appears at the moment to be almost a miracle, so many are the conditions which would have had to have been satisfied to get it going."

Needless to say, the perception that the appearance of life on Earth might have been a freak chemical accident spelled bleak pessimism for the chances of finding life elsewhere. After all, the origin of life is that critical step that marks the transition from an extraterrestrial place being merely "habitable" to it being inhabited. As a result, very few astronomers dared in the 1950s and even the early 1960s to profess belief in the existence of extraterrestrial life in general, and extraterrestrial intelligent life in particular.

Things started to swing in the opposite direction in the late 1960s, first on the chemistry-biology front. Even so, overcoming the conceptual barriers, erected by the conviction that the emergence of life from chemistry was almost inconceivable, required no less than two Nobel Prize–winning discoveries, as well as a complete reversal in our way of thinking about the origin of life.

The first discovery involved the determination of the structure of a specific RNA molecule, the so-called transfer RNA, or tRNA, that is a part of the protein-synthesizing machinery. The complicated three-dimensional figure traced out by the strand of this nucleic acid came as a shock to the scientific community. Quite unlike DNA, with its relatively featureless and rather stiff, repetitive double helix, RNA was found to be a single-stranded molecule, intricately folded up almost like a protein. Robert Holley, a chemist at Cornell University, who was the first researcher to work out tRNA's sequence and 2-D chemical structure, was awarded the Nobel Prize in Physiology or Medicine in 1968, together with Har Gobind Khorana, at the University of Wisconsin, and Marshall Nirenberg, at the National Institutes of Health. A bit later, Aaron Klug of the Medical Research Council in Cambridge and Alexander Rich of MIT determined the surprising 3-D folded architecture of RNA.

A few scientists, including Francis Crick himself and British chemist Leslie Orgel, realized immediately the potential implications of this striking structure—it meant that RNA might be able to act like an enzyme, a biological catalyst, just as proteins do. Orgel then came up with the breakthrough idea that the early life on Earth must have done without DNA and proteins entirely. Instead, he suggested, life started only with RNA! This was a bold speculation at the time, and the notion that RNA might be able to both carry information in its sequence *and* speed up chemical reactions (until then considered in biology to be the exclusive province of protein enzymes) was too much to swallow for most researchers. It wasn't until some twenty years later that, in another dramatic Nobel-winning feat, RNA enzymes were indeed discovered by chemist Thomas Cech and molecular biologist Sidney Altman. This was the seminal step that completely revolutionized thinking about the origin of life. It meant that, in principle, RNA could act as an enzyme to catalyze even *its own replication*, thus potentially solving a thorny "Which came first, the chicken or the egg?" dilemma. All of a sudden, it became possible to imagine a primitive cell that was much simpler than any currently existing cell. In this putative "protocell," RNA molecules played dual roles both as the carriers of genetic information and as the cell's enzymes, performing the basic functions of the cell. The latter included, most importantly, the replication of the genetic information. In this novel scenario, DNA and proteins could be seen as later "inventions" of evolution, custom designed specifically for the tasks of storing information and catalyzing chemical reactions, respectively. The tantalizing conception of a simpler time in the history of life, in which RNA alone played simultaneously all the starring roles in the cast of key cellular actors—being both the "chicken" and the "egg"—became known as the *RNA World*.

On the astronomical side, progress lagged somewhat behind initially, but then things started to advance at breakneck speed. Specifically, on October 6, 1995, astronomers Michel Mayor and Didier Queloz of the University of Geneva announced the first definitive detection of a planet orbiting a Sun-like star outside the solar system. Not surprisingly, they shared the 2019 Nobel Prize in Physics for their groundbreaking discovery.

## Life-Bearing Worlds Galore?

It would be fair to say that on the question of the plurality of inhabited worlds we are now much closer to an answer than thirty years ago, but that the question is still open.

By fall 2023, astronomers had discovered more than 5,500 confirmed extrasolar planets (*exoplanets*) in more than 4,100 planetary systems. More than 930 of these systems have more than one planet. In addition, there were more than 7,400 candidate extrasolar planets, discovered primarily by the Kepler space telescope and the Transiting Exoplanet Survey Satellite (TESS), awaiting definitive confirmation. Can you imagine this? In just about thirty years, astronomy has advanced from a state of not knowing of a single planet orbiting a star other than the Sun, to a treasure trove of thousands and thousands of them! The immediate statistical implication is that our Milky Way galaxy is teeming with planets.

Even more exciting, astrophysicists now estimate that at least one in every five Sun-like or smaller stars in the Milky Way has an approximately Earth-size planet in the star's so-called *habitable zone* (and the occurrence rate could even be as high as one in every three stars or higher). The habitable zone is that "Goldilocks" favorable ring-shaped range of distances from the host star, in which the temperature on the surface of an orbiting Earth-like planet is neither too hot nor too cold, but just right for liquid water (and potentially life) to stably exist. Typically, once the orbit of an Earth-size exoplanet and the properties of the host star (such as its surface temperature, luminosity, and mass) are known, the boundaries of the habitable zone can at least be estimated, assuming a composition for the planet's atmosphere. The atmospheres are generically taken to contain primarily a combination of nitrogen, carbon dioxide, and water vapor, with the last two components assumed to be acting as greenhouse gases. While other factors, such as the atmospheric mass and composition, geological and geochemical drivers, the planet's rotation rate, the presence of nutrients, the availability of an energy source, protection from harmful radiation, and indeed the type and stability of the host star itself, are important in determining whether a planet is truly "habitable," studies suggest that theoretically there could be as many as hundreds of millions of, or maybe even a few billion, habitable planets in the Milky Way.

These staggering astronomical discoveries, coupled with the new, promising chemical-biological insights, have given both the search for extraterrestrial life and the attempts to create life through chemistry in the lab an enormous boost. When these scientific breakthroughs are further combined with existing geological findings on Earth, one may be tempted to conclude that life (of some form) could be ubiquitous. Significantly, geologists have shown that life on Earth was already quite abundant some 3.5 to 3.7 billion years ago—"only" several hundred million years after Earth's surface cooled sufficiently to allow for liquid water to exist. We shouldn't be surprised, therefore, that many have contracted that infectious optimism of the late astronomer Carl Sagan, historically perhaps the most passionate and effective advocate of the search for life elsewhere. Sagan once buoyantly declared: "The origin of life must be a highly probable affair; as soon as conditions permit, up it pops!" A number of biologists agreed at the time. Physiology or Medicine Nobel Prize laureate Christian de Duve went even further to pronounce that the appearance of life in the universe was "a cosmic imperative."

Truth be told, we cannot really be sure of that. There are still many unanswered questions and serious uncertainties on all levels. For example, over the past few decades biologists have been arguing about which of the crucial characteristics of life—being made of cells, metabolism, catalysis, or genetics—emerged first. Predictably perhaps, scientists tended to split into four major camps. There was the "metabolism first" group whose members claimed that the ability to harness resources from the environment to keep the organism alive was the first and foremost capability that had to develop. A second camp argued for genetics or "replication first"—the capacity to generate offspring—undoubtedly a cornerstone of evolution by means of natural selection. A third coterie contended that it is hard to imagine genetics and metabolism without agents that could facilitate and accelerate the rate of chemical reactions, and therefore supported "catalysis first," meaning that protein enzymes had to be a prerequisite for life to emerge. Finally, there was the "compartmentalization first" cohort—those who insisted that life could not have even started without first having some form of a tiny container, a primitive cell, a protocell, to hold together all the molecular principal actors and to separate them from their surrounding environment. Over the years, the members of each of these groups had become so passionately committed to their particular pet choice and so entrenched in their opinion, that at scientific meetings on the origin of life it was not uncommon for attending science reporters to hear a scientist from one camp unabashedly trashing the ideas of all the other groups. Science was almost emulating politics.

Well, that particular problem may have been solved. Amazingly, the most recent findings by origin-of-life researchers seem to suggest that the entire outlook on the origin-of-life question for the past four decades might have been misguided. The "which came first" debate originated from the fact that the leading scenario assumed that one must find a way to construct the first cells one piece at a time, with each component paving the way for the next one. This has changed dramatically in the past few years. Current thinking suggests that one could make the *building blocks* for the subsystems all at once. Researchers have managed to demonstrate that a few simple compounds, which were readily available on the early Earth, could trigger a network of chemical reactions (to be described in detail in the next five chapters) that could have produced—essentially simultaneously—nucleic acids (the core of the genetic molecules), amino acids (from which proteins are made), and lipids (the stuff from which cell walls are constructed). In other words, experiments in Jack Szostak's own laboratory, breakthrough studies in the lab of chemist John Sutherland, and research by many of their colleagues hint that, in spite of being very intricate and precise entities, the first cells might have emerged from a relatively small collection of just the right building blocks. Accordingly, what researchers are now attempting to achieve is more ambitious. Rather than separately examining individual constituents, they try to draw a complete, unified outline—that is, a picture that would successfully integrate all the existing data from lab experiments in *prebiotic chemistry* (the chemistry that preceded life and via which the building blocks of life might have been synthesized) with observations from astrophysics, geology, and atmospheric science, in order to map a robust pathway to life. In this respect, future direct geochemical exploration of Mars (which will be enabled by the return of samples from Mars to Earth) could potentially offer exciting new opportunities. Its findings could bring about a leap in the understanding of life's origins, by allowing us access to an early environment, the type of which has been erased from Earth's geological record, due to recycling by the dynamics of Earth's outer shell.

Of course, neither the spectacular astronomical discoveries nor the promising results achieved so far in the laboratory give a definitive answer to the question of whether life is a freak chemical accident or a cosmic imperative. One could justifiably argue that, in the absence of direct evidence for an uninterrupted chemical route to life, we cannot be sure that even if the conditions are right, the emergence of life is inevitable. Similarly, the fact that astronomers haven't (again, so far) found any convincing signs of extraterrestrial life leaves us in the dark in terms of assessing the probability that such life exists. One cannot reliably calculate the likelihood of an unknown process or an as yet undiscovered phenomenon. British physicist Paul Davies is one of those who correctly points out that just because there are many "habitable" planets in the Milky Way, it doesn't necessarily mean that any one of them (other than Earth) is truly inhabited. We still do not know how likely it is for life to begin, even when an extrasolar planet's temperature and chemistry are propitious. Our Earth's biophilic conditions could have emerged entirely against the odds, and the evolution of an intelligent species could have been an even rarer fluke, rather than a generic outcome of evolution. The existence of humans, in particular, may have been fortuitously facilitated by a series of cosmic contingencies. For example, humans might not have appeared on the scene at all were it not for a serendipitous asteroid impact, some sixty-six million years ago, which led to the extinction of the dinosaurs.

This last point brings about a question that is undeniably as intriguing as the one about the probability of the existence of extraterrestrial life in general. Is there some form of complex or "intelligent" life out there in the Milky Way? In fact, the apparent contradiction between the absence so far of any evidence for the existence of intelligent life, and the expectation that we should have by now seen some signs of a technological civilization (*technosignatures*), has been dubbed the "Fermi Paradox." This designation is based on a famous incident in which the celebrated physicist Enrico Fermi asked a few colleagues: "Where is everybody?"

He was expressing his amazement over the fact that no signs of the existence of other intelligent beings in the Milky Way had been detected. Fermi estimated that under what he considered to be a reasonable set of assumptions, an advanced technological civilization could have reached every corner of our galaxy within a time much shorter than the age of the solar system. The null detection was therefore extremely perplexing. While many potential resolutions to the Fermi Paradox have been suggested over the years, there is still no consensus on which one, if any, is correct. One could even sensibly conclude that the mere fact that there are so many tentative explanations in itself suggests that none of those is truly compelling. More important, however, the Fermi Paradox does raise the unnerving possibility that there may exist some sort of "great filter"—a bottleneck—that makes the *emergence, some stages in evolution, or the long-term survival* of intelligent civilizations exceedingly difficult to transit. This concept was originally introduced by George Mason University economist Robin Hanson in 1996. If true, this could pose demanding implications even for life on Earth. The filter, or probability threshold, could have been in our civilization's past, in which case we may be one of the very few civilizations (or maybe even the first!) to have successfully passed it. This would put an enormous burden of responsibility on our shoulders. But the filter could also be in our future, in which case the COVID-19 pandemic, or the current climate change crisis, may simply represent child's play rehearsals for a future formidable task of successfully surviving such a filter. We shall return to the Fermi Paradox and its ramifications in Chapter 11.

---

We hope that this brief introduction demonstrates that astronomers, planetary scientists, atmospheric scientists, geologists, chemists, and biologists (a large community that includes the two of us) are attempting to solve some daunting puzzles of which we don't have all the pieces yet. Even with the enormous scientific progress we have witnessed in the past few decades, we still don't know whether life is an extremely rare chemical accident, in which case we may be alone in our galaxy, or a chemical inevitability, which would potentially make us part of a huge galactic ensemble. Each one of these prospects entails its own far-reaching scientific, philosophical, practical, and even religious implications. These possibilities may even dictate our course of action with respect to a series of likely existential risks, whether self-inflicted by humanity or of cosmic origin. In some sense, alien life, or the absence thereof, can act as a mirror in which we can examine and contemplate our own accomplishments, but also our culpabilities and shortcomings. Aliens, if they exist, can help us identify and define what it precisely means to be human.

To solve these puzzles, we must take some specific actions. About four centuries ago, Galileo was one of those who gave us a road map for the path we should follow if we want to decipher the cosmos. The only way to find out truths about nature, he contended, is through patient experimentation and careful observations, which can eventually lead to thoughtful theorizing. The theories, in turn, have to be tested by further experiments and observations. This is the basis of the so-called Scientific Method—the somewhat idealized empirical process of acquiring knowledge. As even Sherlock Holmes once noted, "It is a capital mistake to theorize before one has data. Insensibly one begins to twist facts to suit theories, instead of theories to suit facts." We need to continue to simultaneously perform laboratory experiments aimed at finding a chemical pathway to life (if one exists), and astronomical observations with the goal of detecting signs of extraterrestrial life (again, if those are not exceedingly rare). The lab experiments themselves involve two major steps. First, chemists need to fully understand how the building blocks of biology might be synthesized on a young planet. Second, once the right biological molecules exist, biochemists need to discover how a collection of such molecules can assemble to start functioning like a living cell. These findings, in turn, can inform geologists, planetary scientists, atmospheric scientists, and astronomers about the necessary planetary environments that can allow for life to emerge.

As we shall describe in detail later in the book, given the objective difficulties that a quest for life in an immeasurably vast universe (or even just in our home galaxy) entails, and to increase the chances of success, astronomers have adopted a three-pronged plan of attack on this problem. One endeavor concentrates on looking for past or present extraterrestrial life in the solar system. A second effort aims at searching for signs of life (*biosignatures*) in the atmospheres of Earth-like extrasolar planets that are in the habitable zone of their host stars. A third venture attempts to take a shortcut in the entire search process by trying to detect signatures of an intelligent, technological civilization. Here is a brief description of just a few of the existing and near-future life-searching astronomical facilities. With the successful launch of the James Webb Space Telescope (JWST) on Christmas Day 2021, and the preparatory identification of suitable target extrasolar planets for JWST by the Transiting Exoplanet Survey Satellite (TESS), astronomers got their first chance to characterize (or at least detect) the atmospheres of relatively small, rocky exoplanets, or somewhat larger (sub-Neptune) ocean-bearing exoplanets. The researchers' eventual goal will be to search for gases that are far out of chemical equilibrium in a way that could not have been produced by purely abiotic (unrelated to life) processes. As we shall explain in Chapter 9, for example, the discovery of an atmosphere that is very rich in oxygen would suggest a potential candidate for a life-hosting planet, since we know that the oxygen in Earth's atmosphere originated almost entirely from one source only: *life*.

Other exciting projects are also underway. The European Extremely Large Telescope (ELT), a 130-foot-diameter (39 meters) telescope, is planned to start operation in 2028. This telescope, which will be the largest optical/near-infrared "eye on the sky," will attempt to even *image* Earth-like extrasolar planets. Similarly, the Giant Magellan Telescope (GMT), an 83-foot-diameter (25 meters) telescope, is under construction at Las Campanas Observatory in Chile's Atacama Desert, and the Thirty Meter Telescope (TMT) is possibly destined for Mauna Kea in Hawaii. The telescopes are expected to start observing around 2030.

The search for extraterrestrial technosignatures (which started with SETI, the Search for Extraterrestrial Intelligence) has also been gaining momentum. In addition to the Allen Telescope Array, of which the first forty-two elements have been built at the Hat Creek Radio Observatory in rural Northern California, there are projects such as Breakthrough Listen, which targets about a million nearby stars at radio and visible-light wavelengths. At the end of 2019, Breakthrough Listen started a collaboration with TESS, which allows it to scan planets discovered by TESS. China's Five-hundred-meter Aperture Spherical Telescope (FAST) also lists "detecting interstellar communication signals" as part of its science mission. In addition, the Galileo Project is complementary to traditional SETI, in that it searches for physical objects—rather than electromagnetic signals—artifacts that could be associated with extraterrestrial technological equipment.

It would be an exaggeration to claim that we know that the discovery of extraterrestrial life is right around the corner. But all of these endeavors and others do give realistic reasons for optimism. If life in the Milky Way is ubiquitous (or if we simply get very lucky!) we could potentially discover a life-bearing planet within the next one or two decades.

We think that the discovery of extraterrestrial life (especially intelligent life), or the synthetic production of life in the lab, would constitute a revolution that will dwarf the Copernican and Darwinian revolutions combined. What we want to share with you, the reader, is a ringside seat to the captivating quests toward these grand goals. We truly believe that our generation is most likely to play that pivotal role in the history of humankind, in being the first to know where we came from and if we are alone in our galaxy. There is nothing the two of us—the authors—fear more, intellectually speaking, than the thought that we may not be here when these momentous discoveries are made. Perhaps not surprisingly, the inevitability of death only highlights the meaning of the search for life.

There are people, no doubt, who will see the attempts to originate life from chemistry in the laboratory as endeavors to unlock some "forbidden knowledge"—trying to "play God" in some sense. In fact, a poll by the Pew Research Center in November 2021 found that only one in six Americans do not believe in an afterlife, and nearly three-quarters of US adults believe in heaven (which is tantamount to believing that there is more to the origin of life than pure chemistry). We don't feel that investigating the origin of life should be somehow taboo. A strong epistemic curiosity has always driven humans to try to decipher nature's secrets and to answer numerous "How?" "What?" and "Why?" questions. When it comes to something such as *life*—arguably the most precious thing to us as human beings—it would be unconscionable to think that we would not want to find its origins, or to discover whether it is something exclusive to Earth. As Galileo himself once put it: "I do not feel obliged to believe that the same God who has given us our senses, reason, and intelligence, wished us to abandon their use." It is only in what we *do* with the knowledge that we acquire, that we should definitely apply our ethical, moral, and humane principles, to decide what is right and what is wrong.

Some people even argue against endeavors of astronomical exploration and the search for alien life, considering it a dangerous thing to do. Again, while there is indeed no guarantee for the type of relationship humanity may develop with beings that could be dramatically different from us, we do not think that the human sense of wonder, which has always driven efforts far beyond those needed for mere survival, is stoppable.

In his charming book *The Little Prince*, Antoine de Saint-Exupéry describes an inspiring conversation between the narrator and the little prince, before the latter is about to return to his home planet/asteroid. The little prince says: "All men have the stars, but they are not the same things for different people. . . . But all these stars are silent. You—you alone—will have the stars as no one else has them." The narrator wonders: "What are you trying to say?" To which the little prince replies: "In one of the stars I shall be living. In one of them I shall be laughing. . . . It will be as if all the stars were laughing. . . . You—only you—will have stars that can laugh!" Imagine indeed how we would feel if we really knew that a certain extrasolar planet is inhabited, or if we truly understood how life here, on Earth, came into being.

We start our journey of exploration at our home planet Earth. Since life on Earth is the only life-form we know about so far, the first question that chemists have been struggling with is, Could life on Earth truly have emerged from ordinary chemistry? Or, more specifically, could living protocells assemble from chemicals that are expected to have been around on the early Earth? To answer this crucial question, researchers in prebiotic chemistry first embarked on attempts to identify a chemical pathway to the production of the building blocks of RNA and of proteins. The goal of the next step was obvious: to build a cellular system that can undergo Darwinian evolution. We describe these fascinating endeavors, their vicissitudes and successes, and the conceptual revolutions that had to take place, in the next four chapters. Inevitably, there is quite a bit of chemistry involved, and we are aware that many readers may be a bit "rusty" on their biochemistry. We do feel, however, that we have here a unique opportunity to provide interested readers, maybe for the first time, with a truly up-to-date, detailed account of the incredible advances and achievements in this field during the past two decades. We think that the three most intriguing fundamental questions in science are related to *origins*: the origin of the universe, the origin of life, and the origin of mind or consciousness. Of these, the origin of life seems right now to be the most solvable, given current research tools and technologies.

# CHAPTER 2

# The Origin of Life

## The RNA World

*You know life. . . . It's rather like opening a tin of sardines.*
*We are all of us looking for the key.*

—Alan Bennett, *Beyond the Fringe*

Attempts to find a path that leads from the chemistry on the surface of the young Earth to the beginnings of biology encountered many problems right off the bat. First, there was that perplexing issue we mentioned in Chapter 1, of the complexity of modern biology, where everything depends crucially on everything else in a circular way. Recall, for example, that the molecules of DNA and RNA are needed to encode the information that specifies the assembly of those very proteins that are required to make DNA and RNA. This complicating feature introduced obvious "chicken or egg" causality dilemmas. There was, however, a second, even more fundamental problem. That was the question of whether a chemical pathway, in which starting materials are transformed through a series of steps into desirable products, could even exist, without that pathway being shepherded by the enzymes and the control systems of biology. A few researchers have indeed explicitly asserted that the odds of a multiple-step chemical synthesis occurring spontaneously in nature are exceedingly low. Cosmologist and astrobiologist Paul Davies, for example, presented the following probabilistic argument: Suppose that life's origin needs a particular sequence of ten critical and precise chemical steps (he assumed that ten represents, if anything, an underestimate to the number of critical steps that are truly required). Assume further that each one of those steps has a probability of occurrence (during the period throughout which the planet remains habitable) of 1 percent (again, a value he considered to be optimistic). Then the combined probability for life to originate is staggeringly low—one in a hundred billion billion, to be precise.

For many years, these and similar perceived difficulties have been seen as insurmountable obstacles. Impressively, however, origin-of-life researchers now think that they have discovered ways in which Nature could have, in principle at least, managed to solve these types of thorny problems. In this and the subsequent four chapters, we shall follow the remarkable progress that has been achieved in recent years in our understanding of the origin of life. Unavoidably, this brief review involves some of the hardly pronounceable names of compounds usually associated with biochemistry, and a collection of intricate chemical and physical processes. We shall try to concentrate on those parts of the story that have been truly essential along the path of discoveries and breakthroughs. We shall also attempt to highlight the conceptual difficulties that had to be overcome and the ingenious solutions to those impediments. We hope that this approach, even if challenging, will allow for an appreciation of the logic, beauty, brilliance, and patience entailed in the scientific process.

The proposed solution to the first problem—that of the self-referential nature of modern biology—was to posit the existence of a rather different, extremely simple primordial biological cell, or *protocell.* This hypothesis in itself, however, immediately led to new puzzles (in addition to the fundamental question of how such structures came to exist in the first place). In particular, researchers had to understand how these protocells could grow and divide without any of the complex machinery that is available to modern cells. To address this specific hurdle, scientists had to adopt a complete "assumption reversal" process—to take the core notions of the subject, and turn them on their head. This was a bit like what happened in recent years with the taxi industry. Your first assumption if you wanted to start a new cab company might have been that taxi companies have to own cars. The reversal would be that taxi companies don't own any cars. A mere two decades ago, this last concept might have sounded completely unhinged. Today, in contrast, Uber and Lyft are the largest "taxi" companies to have ever existed. The point that origin-of-life researchers had to realize was that whereas modern cells have an *internal* biochemical apparatus that directs growth and cell division (and enables the cells to adjust to a changing planetary environment), most likely, precisely the opposite had to be the case for primordial cells. That is, it was the *environment* that supplied everything in terms of materials and energy to the protocells, and it was *fluctuations in the environment* that provided the engine that effectively controlled cell growth, division, and replication.

To go any deeper into the likely origin and structure of the first cells, we have to consider many additional questions. Those range from geological scenarios and prebiotic chemistry to the very nature of those cells and the evolutionary events that might have led to modern life. Importantly, we should not expect to be able to answer all of these questions at once, and we have to anticipate that there will be quite a few false starts, blind alleys, detours, and setbacks as we attempt to reach a more comprehensive picture. Here is just a partial list of questions we need to answer: What were the key starting materials needed to initiate the process of cell formation? What were the most likely sources of energy that powered the necessary chemical reactions? What were the requirements needed to construct a cozy home for the very first cells? And, perhaps even more important, how many environmental niches were necessary for life to emerge? In other words, did life on Earth need certain environments to generate the building blocks of life, but different ones to nurture life itself once it had started?

On top of these fundamental queries there are many others, some of which are more specific. For example, although the several-decades-old scenario known as the *RNA World*—the stage in the evolution of life on Earth in which self-replicating RNA molecules dominated life processes—provided an attractive view of a simpler time in the history of life, it also raised a host of questions and controversies, many of which are yet to be resolved. The key problem, of course, is figuring out how piles of chemicals that accumulated on the surface of the early Earth could possibly lead to even the simplest of RNA World cells.

There are puzzles on other levels too. For instance, experiments done in the UK Medical Research Council laboratory of chemist John Sutherland, along with work by other colleagues, have taught us much about the chemical pathways that might have led to the emergence of the building blocks of RNA—molecular units known as *ribonucleotides*. But those same experiments have also shown that other, closely related molecules would inevitably be synthesized together with the starting materials of RNA. Unconstrained by the protein enzymes that control the synthesis of everything in modern cells, prebiotic chemistry would have generated a much messier mixture of chemicals. Why then did RNA and not one of those "cousin" molecules materialize from such a muddle? There is also the related, important question: On extrasolar planets, might something other than RNA have emerged as the first genetic molecule of life? Or is there something in the nature of chemistry itself that somehow favors RNA, so that life everywhere in the cosmos has to begin with the very same RNA chemistry? Such wide-ranging questions might at first blush seem to belong to the realm of metaphysics rather than biochemistry, but recent work has demonstrated that a systematic exploration of the chemistry involved can give us compelling answers.

The problem of how to navigate to the RNA World was laid out as a challenge to the scientific community some thirty years ago by chemists Leslie Orgel and Gerald Joyce. The first attempts to address this issue placed in stark relief the question of how to start from the sort of chaotic mixtures that the early origin-of-life efforts seemed to produce in trying to experimentally mimic prebiotic chemistry. This stumbling block—the transition from a confusing jumble to the homogeneous, well-controlled chemistry that we observe in living cells—seemed intractable for many years, but a series of surprising recent discoveries suggests that the solution may be rather simple, almost trivial (in retrospect, of course).

It turns out that a potential answer to at least this key riddle of life, of why RNA and not something else, can be expressed by the unexpected statement: *Because RNA Always Wins!* Here is a brief explanation. Consider our starting point to be a messy "soup" of chemicals, only some of which are the correct seed materials for making RNA. Imagine now that these chemicals are dissolved in a pool of water on the surface of the young Earth, where they are exposed to the intense ultraviolet (UV) light of the young Sun. Amazingly (or perhaps inevitably, depending on your point of view), experiments have shown that the building blocks of RNA tend to be the most resistant to UV irradiation, while many of their cousin molecules are destroyed by UV light. This undoubtedly helps, but we are still left with a fairly complex mixture. The next step toward producing RNA requires the building blocks to join together into chains (to *polymerize*)—essentially creating short, single-stranded bits of genetic materials. While this step has not been sufficiently studied yet, the preliminary evidence suggests that some molecules assemble into chains faster than others. As a result, the less reactive molecules are left behind. Lastly, there is the chemistry of replication itself, in which those small chains are being copied, and the copies are copied again, to produce ever more progeny molecules. Szostak and his colleagues have begun to study this process carefully, systematically comparing the outcomes obtained from different starting materials. The results so far seem to indicate that RNA always wins. The nucleotide building blocks of RNA always react faster than their competitors, so that RNA tends to be created, while the alternatives are constructed more slowly or not at all, and therefore falter. We can think of these three stages—first, resisting UV radiation; second, faster polymerization; and third, more effective copying—as a series of purifying filters. As the original hodgepodge passes through these phases, it becomes progressively distilled, first by UV light, then by chain assembly, and finally by copying chemistry. At the end, a relatively homogeneous RNA emerges, clean and ready to achieve its destiny of giving birth to the RNA World.

We don't want to give the readers the impression that this story of how RNA might have triumphed over its rivals, and emerged as a champion to begin life and dominate its evolution, is without its critics or free of controversy. Indeed, there is a vigorous discussion about all aspects of this narrative. Whether or not it is really the case that, out of all the myriad possibilities, only RNA has the right properties to initiate life is a very complex question, to which we are not likely to know the definitive answer for quite some time. Whereas there is no doubt that a systematic synthesis and examination of alternatives will rule out many relatives of RNA, this approach will always leave us wondering whether there is something else that is just as adequate as RNA, which we have simply not considered yet. What could give us an answer (at least in principle)? The most convincing evidence, of course, would come from the discovery of life on some distant world (such that we could be certain that it had evolved independently of life on Earth). But even that would not give an immediate answer. The first step would indeed be to find compelling signs for life on other planets. That discovery, if and when it comes, would at least demonstrate that life is not incredibly hard to start—there is no impassable bottleneck. We would instantaneously know at that point that we should expect that a relatively simple pathway from chemistry to the beginnings of life exists, in which each step has a reasonably high probability of success. Even so, finding out whether life on exoplanets also started with RNA will remain a huge challenge, unless that alien life includes intelligent beings who are willing to communicate with us.

## Looking Back from Modern Life into the Past: The RNA World

At the beginning of this chapter, we described how the confounding complexity of modern life erected a conceptual barrier, which for many years obstructed reasoned thinking about the origin of life. The recognition that very early life had to be extremely simple, with RNA playing a central role as both a means of storing information (although not as robustly as DNA) and the molecular basis of the first catalyzing enzymes (even though RNA is not as good a catalyst as protein enzymes), provided researchers with a fresh insight, and allowed for a simplifying breakthrough. In the late 1960s, three scientists were the first to realize the significance of what has become known as the RNA World: Carl Woese, now famous for his work on the evolutionary tree of life; Francis Crick, of structure-of-DNA fame; and Leslie Orgel, one of the true pioneers of prebiotic chemistry (as mentioned in Chapter 1). All three perceived that the fact that RNA chains could fold up into complicated three-dimensional shapes implied that RNA might be able to act as an enzyme—could catalyze chemical reactions—just like proteins. The ramifications of this understanding were astounding: if RNA could catalyze its own synthesis, the origin of life might come down simply to the origin of a self-replicating RNA, or an *RNA replicase* (an enzyme that catalyzes the replication of RNA from an RNA template). Unfortunately, with the attention of the scientific community at the time having been focused on unraveling the mysteries of protein enzymes, no one took seriously the idea that RNA could act as an enzyme, and this crucial key to the origin of life languished in obscurity for some fifteen years.

The news that RNA molecules could act as enzymes eventually struck the scientific community like a thunderbolt only in 1982. In that year, two separate groups of scientists discovered RNA enzymes hiding in plain sight, within two very different parts of modern biology. Tom Cech, a biochemist at the University of Colorado, Boulder, had been studying the process of RNA splicing for several years. RNA splicing is in itself a somewhat puzzling process, in which cells copy the information stored in DNA into long RNA chains, and then mysteriously clip out and discard chunks of that chain by cutting it twice in the middle and joining the ends back together again. RNA splicing is widespread in biology, but how exactly it occurs was unknown in the early 1980s, and many labs were racing to uncover the underlying mechanism. Tom Cech decided to study splicing in an esoteric microorganism with the rather complicated name *Tetrahymena thermophila*—a ciliated unicellular organism that is commonly found swimming around in small ponds. This organism had the convenient property of making a very large amount of a particular RNA, which it then spliced in a fairly simple manner, thereby making it an ideal system in which to study how splicing worked. At the time, the general assumption was that the process of splicing was carried out by protein enzymes, much like all other known chemical reactions in cells. Consistent with this hypothesis, Cech set out to purify the protein or proteins responsible for splicing by first purifying the unspliced RNA, and then adding back cellular proteins in the hope of seeing splicing as it was taking place. Frustratingly for him, however, he was unable to separate the splicing activity from the RNA itself. After strenuous, multiple unsuccessful efforts, he felt compelled to conclude that the RNA must have been catalyzing its own splicing.

Needless to say, this conclusion was received with some skepticism by the scientific community, which was still wedded to the idea that all enzymes are proteins. Critics even went so far as to claim that Cech must have simply failed to remove the catalytic protein from his RNA preparation. This incredulous response inspired Cech to do things differently. He obtained unspliced RNA not from *Tetrahymena* cells, a process that might have led to an inadvertent contamination with the long-sought splicing enzyme, but by making the unspliced RNA in a test tube, from DNA and just one bacterial enzyme that could transcribe the DNA into RNA. What he found was amazing: while RNA prepared in this way could not possibly contain any splicing enzyme, it still spliced, all by itself! In other words, in this roundabout way, a fruitless quest to purify a protein that turned out not to even exist opened up a new exciting window on biology—the discovery of RNA enzymes, also known as *ribozymes*.

That was not the end of this story. In one of those astonishing "when the time is right" coincidences, at the same time that Cech was unsuccessfully trying to purify his splicing enzyme, Yale University molecular biologist Sidney Altman and his colleagues were studying an RNA-processing enzyme known as *Ribonuclease P* (*RNase P* for short). This enzyme cuts certain cellular RNAs in a very specific manner, and Altman had found that the enzyme consisted partly of RNA and partly of protein. Again, the initial assumption was that the protein component was doing all the actual work, while the RNA component was playing an assisting role, perhaps by recognizing the RNAs to be cut by the protein enzyme. In this process, the protein turned out to possess a very large positive electric charge, which made sense, since it had to bind the highly negatively charged RNA component of the enzyme, which in turn had to bind the negatively charged RNA substrate that it had to cut. This finding (of the large positive charge) led Altman to the non-mainstream idea that maybe the protein was nothing but a passive bystander, whose role was simply to stabilize the RNA complex by neutralizing the large amount of negative charge. If so, he reasoned, perhaps the positive charges could also be supplied in an entirely different way. Indeed, Altman and his colleagues discovered that by adding enough magnesium ions (each one of which carries two positive charges) to the RNA component of RNase P, enzymatic activity could be seen without any added protein. As is often the case in science, this second example of a ribozyme was soon thereafter followed by a flurry of discoveries of small self-cleaving RNAs, firmly solidifying the idea that RNA molecules can truly catalyze chemical reactions.

The discovery that RNA molecules could act as enzymes completely revolutionized thinking about the origin of life, and the importance of this discovery was highlighted by the 1989 Nobel Prize in Chemistry, awarded to Cech and Altman. Suddenly, the earlier ideas of Crick, Orgel, and Woese about the centrality of RNA seemed obvious. With one simplifying realization, it was no longer necessary to imagine some complicated scheme through which RNA and proteins could arise together. Rather, an earlier, simpler form of life could be envisaged, in which RNA molecules played dual roles as both the carriers of hereditary information and the catalysts of key cellular biochemical reactions. It was this idea of a prior form of life in which the main molecular player was RNA that was popularized by Harvard biochemist Walter "Wally" Gilbert in the pithy catchphrase "the RNA World."

Of all the hypothetical RNA-catalyzed chemical reactions of the RNA World, arguably the most important one was the replication of the cellular RNA genome itself. This could be imagined to be the job of a presumptive ribozyme (RNA enzyme), which we referred to earlier as an RNA replicase. This almost magical RNA would have been a special RNA sequence that could copy itself, thereby igniting the exponential replication that is one of the hallmarks of life. Due to its pivotal role in the origin of life, the dream of generating an RNA replicase in the lab has been pursued by laboratories around the world.

Although the discovery of ribozymes was vital to the formulation of the RNA World hypothesis, there were other clues that pointed to an earlier RNA-centric stage of life. One of these clues came from a puzzling aspect of cellular metabolism, long seen as an enigma but now elevated to the status of crucial evidence for the nature of early life. All modern cells use protein enzymes to catalyze (almost) all the myriad chemical reactions that constitute cellular metabolism. However, hundreds of these enzymes cannot do their jobs unassisted. Rather, they require the help of smaller molecules referred to as *cofactors*. Interestingly, many (but not all) of these cofactors are composed of two constituents, one of which is the chemical that helps the enzyme to speed up a chemical reaction. The other is a nucleotide—one of the building blocks of RNA. Why would so many distinct cofactors have a tiny piece of RNA as part of their structure? This fact made no sense until the RNA World hypothesis provided a potential explanation—these baffling molecules could be remnants, "fossils" in a sense, of the RNA World. Perhaps at a time when RNA was struggling to catalyze cellular metabolism, these RNAs got a chemical boost from small cofactors that could diversify the chemical repertoire of RNA with additional chemical groups. If these cofactors were attached to the beginning or end of an RNA chain, they would be conveniently placed to aid in catalysis. Later on in the evolution of life, one could imagine ribozymes being replaced over time bit by bit by protein enzymes, with the RNA component gradually shrinking and the protein component growing, until all that was left of the RNA and its cofactor was what we see today—a bizarre-looking cofactor that is half warhead and half RNA.

Modern cells hold even more hints as to their ancient past, and one of those is now seen as the ultimate "smoking gun" evidence for the reality of the long-lost RNA World. To understand this important indicator, we have to examine the way in which proteins are produced inside all existing living cells. While the process itself is quite complicated, the key point is not to get lost in the details and to appreciate the crux of the matter.

Let's look first at how the information used to direct the synthesis of a particular protein is transmitted and decoded. Protein production starts with *transcription* (from DNA to RNA) and continues with *translation* (from RNA to protein). The information is stored in the particular sequence of the bases in DNA (in cells, or in RNA in some viruses). That is, each building block (nucleotide) of DNA contains one of four nitrogen-containing chemical bases (whose names are often denoted by the abbreviations A, T, C, and G), and the coding of the genetic instructions that are needed to create a given protein is contained in the specific order of those four letters in the sequence. In the double-helix structure of DNA, C always connects (or pairs) with G, and A always pairs with T, to form something resembling rungs in a ladder. The first step in gene expression is for this coded information to be transcribed into a single-stranded molecule of RNA known as *messenger RNA*, or *mRNA* for short. (Incidentally, messenger RNA became widely known as the key component of the vaccine against COVID-19.) When an mRNA is transcribed from DNA, the mRNA contains within it the sequence of bases that codes for a particular protein, but that mRNA sequence has to be decoded so that it can be translated into the sequence of *amino acids*—the building blocks of proteins—in a protein chain. For instance, the DNA sequence GCT results in an mRNA sequence that codes for the amino acid alanine; the genetic code relates three nucleotide sequences to one of the twenty amino acids. The entire process involves multiple other RNAs (in itself a clue, if you think about it, to RNA's central role), and the largest of those RNAs are the RNA components of the *ribosome*—the molecular machine responsible for the synthesis of all coded proteins in every cell of every organism on Earth. The ribosome is an enormous molecular apparatus with an extremely ancient evolutionary history. The ribosomes of different organisms have closely related structures, and their ribosomal RNAs are related as well, pointing to a single common origin.

But why does the ribosome have these large RNA components? For many years, the ribosomal RNAs (or rRNAs for short) were viewed as a kind of passive scaffold whose job was to organize and position the large number of proteins that make up the remainder of the ribosomal structure (much as the RNA component of RNase P was initially viewed as a passive supporting entity). That viewpoint gradually started to change as our biochemical and structural understanding of the ribosome evolved. Here is how the ribosome decodes the information in an mRNA to direct protein synthesis. The ribosome has two "halves," a "small half" and a "big half," referred to as the small and large subunits. The decoding process itself is done by the small subunit. It holds the mRNA in a particular way, such that it is kinked right between the last unit of genetic code to have been translated and the next one to be translated (the units of the genetic code themselves are called *codons*). This kink allows the two codons to be recognized by two small RNA molecules known as transfer RNAs (tRNAs), and those act as adaptors and bind the codons with their complementary sequences of nucleotides. Ultimately, this molecular recognition is what brings the correct tRNAs together in the right order. Far away from this site, joined to the ends of the tRNA molecules, lie the amino acids that will, in turn, become joined together. These amino acids are held close together and in the right orientation to react, within the center of the "big half"—the large subunit of the ribosome. This is the real "enzyme" that eventually makes proteins, and remarkably, this site is composed entirely of RNA. That is, the "enzyme" is in fact an RNA enzyme. In the words of Yale University biochemist Thomas Steitz, *"The ribosome is a ribozyme."*

Even if you found all of these unfamiliar biochemical steps somewhat dizzying, the bottom line is very simple: *the RNA components of the ribosome are not just passive bystanders; they are in fact the very molecules that catalyze the synthesis of all of our proteins!* The implications of these startling findings are clear: since RNAs make proteins, RNAs must have come first. This is the "smoking gun" evidence that clinches the RNA World hypothesis of an earlier, simpler time, prior to the evolution of modern synthesis. A time when enzymes were made of RNA.

Two other aspects of modern cellular metabolism also support the idea of the ancient primacy of RNA. In all modern cells, genomic information is archived in DNA. Astonishingly, and for no obvious reason, the building blocks of DNA (*deoxynucleotides*) are synthesized in cells by modifying *ribonucleotides*, the building blocks of RNA. Why would this be the case? One attractive explanation might be that primordial cells did not contain DNA, and thus only needed to make ribonucleotides for the synthesis of RNA. Later, as cells evolved to use DNA, the simplest way to make DNA might have been through transforming ribonucleotides into deoxynucleotides.

Finally, the multitude of roles played by RNA in all modern cells provides in itself circumstantial evidence for the primordial origins and initial dominance of RNA. For example, in bacteria, RNAs known as *riboswitches* regulate numerous metabolic activities, while in eukaryotes (organisms whose cells contain a nucleus; humans are in this category), other types of non-coding RNAs regulate gene expression (the process by which the information encoded in a gene is turned into a function). That is, regulatory RNAs can control which genes are expressed in a particular cell, in turn determining what the cell can do. This is achieved by controlling the stability of mRNAs and their ability to be translated. Again, the simplest explanation for the multiple roles of RNA is that life developed with RNA as its genetic material and it also used RNA for catalysis and for regulatory activities. At later times, as life evolved, the role of RNA in information storage was supplanted by DNA, a much more chemically stable molecule whose robustness makes it a better choice for archiving valuable information. Similarly, the role of RNA in catalyzing chemical reactions was largely superseded by protein enzymes, which are more effective catalysts because they possess a greater diversity of chemical groups.

Realizing that RNA was key to life's origins, and having seen compelling evidence for the reality of the RNA World scenario, we are now ready for the next step in our attempt to understand the emergence of life—exploring whether there is a natural pathway for the chemical production of the building blocks of RNA.

# CHAPTER 3

# The Origin of Life

## From Chemistry to Biology

*To raise new questions, new possibilities, to regard old problems from a new angle requires a creative imagination and marks the real advances in science.*

—Albert Einstein and Leopold Infeld, *The Evolution of Physics*

*The fun in science lies not in discovering facts, but in discovering new ways of thinking about them.*

—Sir Lawrence Bragg, *A Short History of Science*

We have had to undergo a huge conceptual inversion from considering the complexity of modern life to thinking about the simplicity of primordial life, with RNA as its single biological polymer. But this still leaves us with the not insignificant problem of how we go from a messy collection of chemicals on the early Earth's surface to the organized structure of the first living cell. It isn't even obvious that this is a problem that can be solved. Some might argue that this is not a legitimate subject for scientific inquiry, since we cannot go back in time to observe what truly happened, and therefore no hypothesis can be genuinely tested. These objections, however, are overly pessimistic, since we can certainly develop hypothetical scenarios that are falsifiable, in the sense that proposed pathways must be chemically realistic, have to proceed under geologically reasonable environmental conditions, and should be self-consistent in terms of leading stepwise from abundant starting materials and sources of energy to the more complex chemicals needed to build a simple cell. Trajectories and processes that degenerate into mixtures of millions of compounds, or end up in useless intractable polymers (such as kerogen or tar), can be ruled out, and should direct our attention elsewhere. Given the relative complexity of chemical structures and reactions, in the following, we include a small number of chemical diagrams, which we hope will help to visualize the molecular rearrangements and processes involved. At the end of this chapter, we have added an Appendix to clarify how to read such diagrams.

The critical question is whether we can trace the outlines of productive routes from simple initial materials to the central compounds of biology. To begin, we have to identify the necessary feedstocks and energy sources, but we also need to know where we're going, that is, what we need to get biology started. The key compounds for life on Earth are mostly composed of carbon, nitrogen, oxygen, and hydrogen, with smaller proportions of phosphorus and sulfur. Since hydrogen is ubiquitous in the universe and in chemistry, we won't explicitly worry about hydrogen in what follows (and in the illustrations of chemical structures we don't even show most of the hydrogen atoms—please see Appendix).

To build RNA, which, as we have argued, was necessary for the first cells, we need to make its building blocks—the nucleotides—which are themselves quite complex chemical compounds. Nucleotides consist of three pieces, a nucleobase (the information-bearing chemical

The nucleotide 5′-AMP, with the adenine nucleobase at right, the ribose sugar in the center, and the phosphate at left.

unit), a sugar (ribose in the case of RNA), and a phosphate group (that links the nucleotides together in a chain).

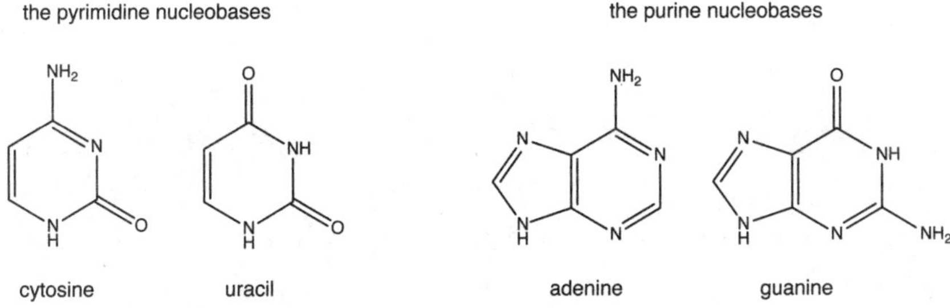


In RNA there are four nucleobases, denoted by the abbreviations (the first letters of their names) A, G, C, and U, and they are composed of carbon, nitrogen, and oxygen (as we have seen, the T in DNA is replaced by U in RNA). The sugar is composed of carbon and oxygen, and the phosphate group consists of phosphorus and oxygen. To produce the nucleobases, a starting material that contains both carbon and nitrogen would be ideal, and indeed it was recognized over fifty years ago that *adenine* (the nucleobase abbreviated as A) is nothing but five molecules of the extremely poisonous and flammable hydrogen cyanide (chemically HCN) joined together in a very specific way. In a series of classic experiments during the 1959–1962 period, University of Houston biochemist Joan Oró boiled a solution of HCN (very carefully!) and obtained, among other compounds, adenine.

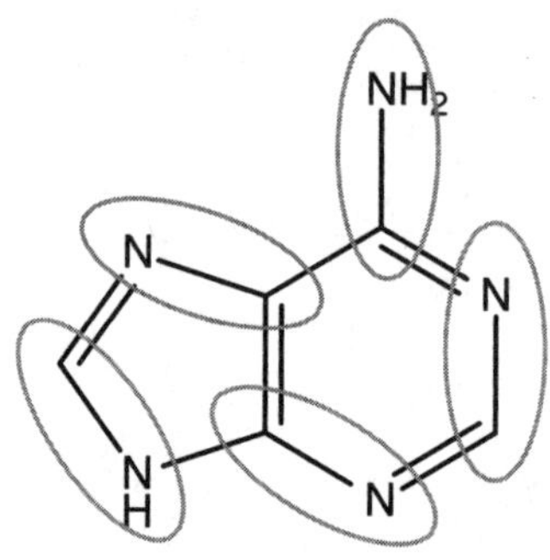


In the above figure of the structure of adenine, pairs of carbon and nitrogen atoms are enclosed in ovals, with each such pair representing one cyanide unit. Those experiments sparked optimism that simple paths to all of the remaining nucleobases and the corresponding nucleotides would soon be worked out. Alas, this was not to be. In experiments aimed at generating both A and G (the so-called *purine* bases), only traces of G were obtained, along with many other related compounds that are not part of the makeup of RNA. The two remaining nucleobases, C and U, seemed as if they would be simpler to deal with, especially since U can be obtained from C by a reaction with water. Notably, the nucleobase C could be obtained by a reaction of two simpler compounds, both of which could arguably have been present in reasonably high concentrations in local environments on the early Earth. The first of these is urea, a common metabolite in modern biology, and famous as the first organic compound ever made in the laboratory. Recall that in the late 1820s the German chemist Friedrich Wöhler made urea simply by heating ammonium cyanate (itself a derivative of cyanide). Interestingly, urea is also the product of the reaction with water of another cyanide derivative called cyanamide, that can in turn be generated in many ways. For example, cyanamide can be produced in reducing atmospheres (those lacking oxygen and other oxidizing gases) that are rich in hydrogen and are exposed to ultraviolet light. (This compound has been detected in the atmosphere of Saturn's moon Titan, which, as we shall see in Chapter 8, is one of the targets in the search for potential extraterrestrial life.) The other starting material needed to make C is a compound with the more complicated name of *cyanoacetaldehyde*, which is the product of the reaction of *cyanoacetylene* (an organic compound that was also detected in Titan's atmosphere) with water. Under the right laboratory conditions, very concentrated urea and cyanoacetaldehyde efficiently combine to generate the C nucleobase, as shown by Leslie Orgel and Stanley Miller in the 1970s (although these two chemists engaged in vigorous arguments about the plausibility of such a synthesis).

At this point, the picture of the origin of life looked promising, since it seemed that key nucleobases of biology might be (relatively) easy to make under prebiotic conditions. Unfortunately, a closer examination of the next steps revealed unexpected difficulties. The sticking point was that having the nucleobases is not enough, since they have to be connected to the sugar ribose in order to make RNA, and it turned out that this particular reaction just does not work. In biology, the reaction is catalyzed by enzymes, and the process uses the sugar ribose with phosphate groups attached to it in specific places. Neither one of those requirements seems plausible in known prebiotic chemistry. In fact, frustratingly, even just making the sugar (ribose) itself presents a severe problem.

At first, making ribose looked simple, since it can be produced by heating the simple organic compound formaldehyde ($CH_2O$, itself thought to be abundant in the early Earth's atmosphere) in water, with a bit of calcium hydroxide (often called slaked lime).

This process results in a complicated set of reactions called the *formose* synthesis, which in essence turns the poisonous formaldehyde into sweet-tasting sugars. Curiously, in the same way that adenine can be made from five cyanide molecules, ribose consists of five formaldehyde

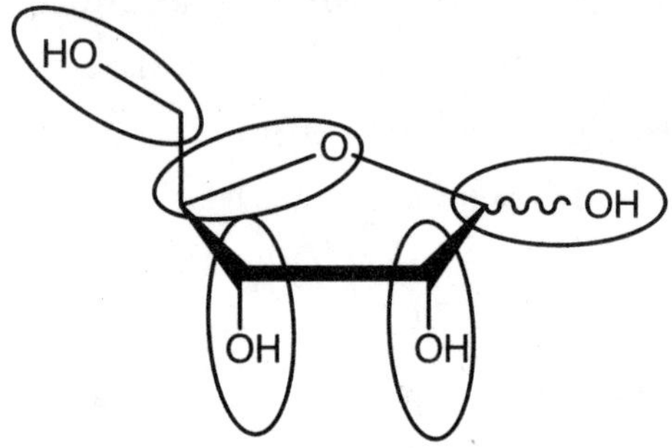


Ribose, viewed as five formaldehyde units stitched together.

units assembled into a ring. In the above figure, each oval encloses a pair of linked carbon and oxygen atoms derived from one formaldehyde unit. That sounds promising, except that the ribose specifically needed for RNA construction is typically less than 1 percent of the complex mixture of sugars that is formed. Moreover, subsequent reactions create numerous messy products that eventually turn everything into a useless tar. Nevertheless, because of its simplicity and especially because of its autocatalytic nature (i.e., the reaction is catalyzed by one of its products), the formose reaction continues to be studied as a simple means of making sugars from an abundant feedstock. A variety of approaches have been explored in an effort to "tame" the formose reaction, such as conducting it in the presence of borate salts. While borate is a common mineral in some geological settings, its relevance in prebiotic scenarios remains uncertain. In addition, even though borate does simplify the mixture of products generated by the formose reaction, the reaction network still remains complex and many sugars are generated. In an interesting new approach, experiments in which the formose reaction is carried out in the presence of other likely feedstock molecules (such as cyanide and cyanamide) are just beginning to be explored. For the time being, therefore, let's put the formose reaction aside, while keeping in mind that it is an alternative potential source of sugars.

As we have already seen a few times in other contexts, finding a solution to the convoluted problems of nucleotide synthesis required a conceptual revolution. In this particular case, it required several such revolutions. The primary psychological roadblock to progress came from the way we instinctively perceive the chemical structure of nucleotides. It is tempting to mentally separate that structure into three parts—a nucleobase (carbon and nitrogen), a sugar (carbon and oxygen), and a phosphate (phosphorus and oxygen). Consequently, for chemists it was only natural to imagine these components as being made separately, and then combined stepwise to generate first a *nucleoside* (a nucleobase plus sugar) and then a nucleotide (adding a phosphate at the end). In practice, carrying out cyanide chemistry in the presence of formaldehyde immediately results in the synthesis of a product called a *cyanohydrin* (formed by the rapid reaction of cyanide with formaldehyde), which for a long time was thought to be a useless dead-end product. This intuitive combination of assumptions on the one hand and the correct, but misleading, chemistry on the other worked in tandem to block further progress for decades.

The first attempt to overcome this impediment was extremely creative, even though initially it did not seem particularly encouraging. The idea was to break away from the concept of making the nucleobase and sugar separately (and then combining them to make a nucleoside). Instead, the new thinking was to make an earlier intermediate compound, one that could subsequently be transformed into the prized nucleoside. The first baby step in this direction was taken by Orgel, who showed that ribose reacted very cleanly with cyanamide, a close relative of cyanide that we have already encountered as a possible precursor to the nucleobase cytosine, or C. Remarkably, the reaction of cyanamide with ribose forms a beautiful, crystalline compound with the somewhat unwieldy name of *ribose aminooxazoline*, which we'll refer to as *RAO* for short. The fact that RAO crystallizes out of the reaction mixture provides a stunning advantage to this approach, since one can imagine a reservoir of RAO slowly accumulating, building up over time, with by-products being washed away and thereby resulting in a natural purification process. This scheme, of building up a reservoir of a purified intermediate compound greatly reduces the apparent difficulty of generating a complicated compound such as a nucleotide, where a series of reactions must occur in a defined order. As we'll see, this type of process occurs again and again in prebiotic chemistry, and it is one of the key concepts supporting the plausibility of the prebiotic synthesis of the building blocks of biology.

Setting aside for the moment the problematic aspect of making RAO from ribose, the question is how we would go from RAO to the desired nucleoside. As it turns out, two simple steps can get us close to the nucleoside C, but there is still a hitch, because the version of C that we obtain is not quite the same as the form of C used in biology. In the type of C made from RAO, the C nucleobase is pointing down from the sugar, not up. Technically, the biological version is referred to as the β-anomer (a kind of isomer—same atoms but a different structure), while the version obtained from RAO is called the α-anomer. Can we get around this additional problem? Exposure to UV light does cause a small, tantalizing conversion to the biological form of C, but with a very poor yield of only about 4 percent. At this point we can see at least two major obstacles with this hypothetical pathway to making C: first, beginning the synthesis of RAO from pure ribose is not realistic in the absence of a way to purify and store this unstable sugar, and second, we end up with the wrong anomer of C and we have no way to convert the α-anomer to the needed β-anomer. There the matter lay for about twenty years, until the early 1990s when the British chemist John Sutherland decided to rethink the entire process of nucleotide synthesis.

In a series of papers published over fifteen years, Sutherland and his team slowly inched closer to a solution, culminating in a 2009 paper that marks a true turning point in the study of prebiotic chemistry.

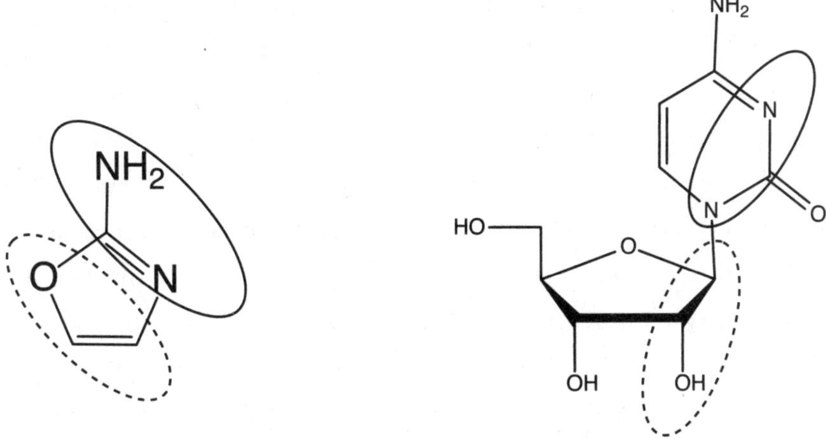


Relationship between 2AO (on left) and the nucleoside C (on right). The atoms of 2AO that become part of the ribose sugar of C, as shown enclosed in the dashed ovals. The atoms of 2AO that become part of the nucleobase of C, as shown enclosed in the solid ovals.

First, Sutherland pushed back the beginning of the synthesis, so that starting materials simpler than the sugar ribose could be used.

Accordingly, in the initial step of their process, Sutherland and his colleagues allowed the simplest possible sugar (*glycolaldehyde*), which has only two carbon atoms, to react with cyanamide (the same cyanide-related compound used above to make RAO). These two compounds react with each other to make the simple ring-shaped molecule *2-aminooxazole*, which we'll call *2AO* for short. The advantage of this step is that a new carbon-nitrogen bond is created, and this is precisely the bond that at the end of the pathway will join the sugar to the nucleobase (to produce the desired building block nucleoside). The important point is that the very C-N bond that previously just wouldn't form by direct reaction of the sugar with the nucleobase is now introduced right at the beginning of the pathway, where it forms readily. The compound 2AO was in fact already well known, so it is interesting to understand why it had not been previously considered as a good intermediate on the way to making the C nucleobase. The answer comes from the historical traditions of organic chemistry, where reactions tend (or tended, since times have indeed changed) to be considered in isolation. When the simple sugar glycolaldehyde and cyanamide are mixed and are allowed to react with each other in isolation, very little 2AO is produced, and most of the material ends up as a complex tarry mess. Upon considering the details of the reaction mechanism, however, Sutherland concluded that this problem might be solved by the presence of a buffer that would keep the acidity of the reaction roughly constant, and also by a common type of catalyst—basically a molecule that can help to shuffle protons around more easily. Remarkably, phosphate is pretty good at doing both of these jobs. Crucially, we also know that phosphate had to be available for this prebiotic chemistry, since phosphate is part of the structure of nucleotides and RNA. So, Sutherland's group added phosphate to their reaction mixture, and lo and behold, a reaction that previously yielded only traces of the desired product now became highly specific and efficient. This simple step of adding another component to a reaction, even though it is not present in the final product, became one of the iconic examples of *systems chemistry*—the idea of using materials that absolutely had to be around (just as phosphate must have been there since it is a part of nucleotides and RNA). We should still emphasize that this step was not at all obvious, because the received wisdom of the time was that phosphate would only be available in trace concentrations due to its precipitation in the presence of calcium as the mineral *apatite.* Using high concentrations of phosphate was a leap of faith that in itself brought another problem, the so-called phosphate problem, into high relief. Putting aside that issue for the moment, we still have to contend with the fact that 2AO is a very small and simple molecule that seems rather far removed from the C nucleoside that is ultimately the goal of our inquiry.

It turns out that progressing from the first intermediate, 2AO, to the next intermediate, RAO, is on the one hand straightforward, but on the other problematic. The good news is that 2AO reacts rapidly with a key three-carbon sugar (*glyceraldehyde*), to produce RAO, which, as before, could crystallize out of the reaction mixture for a very natural purification process. In addition, a major side product called AAO is generated, which remains in solution. However, this step required a second leap of faith, since it created the new problem of the synthesis of the three-carbon sugar itself, and the even harder problem of how the two-carbon sugar (and *only* the two-carbon sugar) could be used in the first step and the three-carbon sugar (and *only* the three-carbon sugar) in the second step. Worse yet, the relevant three-carbon sugar is unstable, and it relatively rapidly converts to a configuration of the atoms (an isomer) that would not yield RAO. Finally, we are once again stuck at the stage of RAO, which, as we had previously found, gives us the wrong (non-biological) form of C.

Are we then as much at a dead end as we were previously? Not quite, because Sutherland had one additional card to play, by switching the focus to the other product of the reaction of 2AO with glyceraldehyde, namely that side product AAO. This compound had been neglected in the past because it involves a sugar that, while a close relative of ribose, is yet different enough that it cannot form a genetic polymer like RNA. This sugar, arabinose, differs from ribose only in that the oxygen atom attached to the 2′-carbon is above the sugar ring, not below it as in ribose. Why did Sutherland decide at this point to go with AAO instead of RAO, given this problem? Remarkably, it was the phosphate that once again came to the rescue. After the reaction of AAO with cyanoacetylene (the step that builds the nucleobase C), Sutherland was able to attach a phosphate to the sugar in such a way that the phosphate could attack the adjacent carbon atom and in one almost magical step generate a proper C nucleoside (a diagram of this reaction is shown in section 6 of the Appendix to this chapter). The principle behind this critical step is the *principle of intra-molecular reactivity*, which is just a fancy way of saying that two chemical groups are more likely to react with each other if they are held close together.

Fascinatingly therefore, in the seminal 2009 paper by Sutherland and his colleagues, we see the operation of three important concepts:

*systems chemistry* (in the use of phosphate to make 2AO), *intra-molecular reactivity* (in the phosphate attacking the atom next door), and, perhaps just as important, *a judicious deferral of unsolved problems until a later time* (for example, the source and timing of glyceraldehyde addition). Let us now reconsider that rather long list of deferred problems, and see how the pathway to the key building blocks of RNA has been gradually simplified since that breakthrough paper. In so doing, we shall see a picture slowly emerging as to how this intricate series of chemical reactions could have realistically happened on the early Earth.

## From the Lab to Nature

The important problem researchers have been facing was, and still is, that of turning a chain of chemical reactions from being a mere laboratory demonstration into something relevant to early Earth conditions. In the laboratory, chemical reactions are usually examined one at a time, and intermediates are purified before going on to the next step. There are many sophisticated ways of purifying compounds in the lab, but the question is whether there is anything analogous in the natural world. Fortunately, a moment's reflection reveals that of course there is—the precipitation or crystallization of pure minerals in geology is common, indeed universal. Beautiful crystals of everything from common quartz to rare and exotic minerals can be seen in any museum of natural history. On the modern Earth, crystals of organic compounds are extremely rare. The reason is, at least in part, that organic compounds provide food for microorganisms and tend to be rapidly consumed by bacteria and fungi. In addition, the natural chemistry of the modern world is completely different from that of the early Earth—think of our oxygen-rich atmosphere, and contrast that with the oxygen-free conditions of our young planet. Can we then think of ways in which certain key compounds might precipitate (be deposited in solid form) or crystallize out, effectively building up reservoirs of purified materials that could accumulate over time, even being purified as groundwater filters through the crystalline mass, washing away impurities? Indeed, there are several very interesting compounds with the potential of behaving precisely in this manner. We shall now consider those, starting from the simplest (cyanide) and progressing to the most complex (RAO).

Since cyanide has long been considered one of the most likely starting materials for the prebiotic synthesis of biomolecules, let's have a closer look at how cyanide might have been made and subsequently stored in a concentrated form, suitable for synthetic reactions. One of the properties that makes cyanide so potent is that there is a considerable amount of energy stored in the triple bond linking its carbon and nitrogen atoms. What that means is that cyanide is in a sense primed to react with other molecules, and those transformations can proceed energetically downhill. This avoids the complexity of having to inject more energy into the system to drive the desired reactions. However, the same property also creates a problem, because cyanide reacts (albeit slowly) with water. This so-called hydrolysis reaction (the chemical breakdown of compounds by water) degrades our precious cyanide into a less reactive product called *formamide*, which although interesting in its own right (it is a liquid at moderate temperatures, and is excellent at dissolving some molecules that are almost insoluble in water) is less useful as a synthetic building block. Moreover, formamide also reacts slowly with water, this time generating ammonia and formic acid. The problem then is how to rescue cyanide from its seemingly inevitable destruction by water. This is a truly bothersome issue because cyanide was most likely formed in the atmosphere, where its concentration would have been quite low. Atmospheric cyanide would have been brought to the surface after dissolving in raindrops, but the resulting rainout of cyanide would again be expected to be in the form of a very dilute solution in water. For many years it was thought that this dilute, dissolved cyanide was essentially a dead end due to its inevitable hydrolysis.

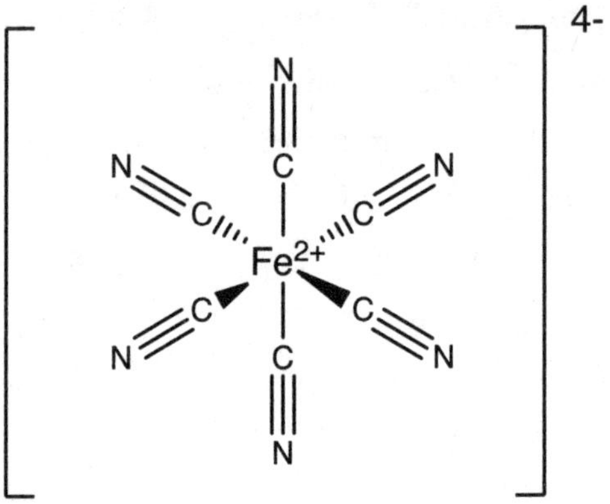

As is so often the case, a potential solution to this cyanide problem was common knowledge in the field of chemistry, but no one had applied that wisdom to solving this critical issue in prebiotic chemistry.

The solution to the apparent difficulty in concentrating cyanide emerges from the extremely strong and fast interaction of dissolved cyanide with certain metal ions. Most noteworthy in this respect is the fact that dissolved ferrous iron (an iron atom that has lost its two most weakly bound electrons) reacts with cyanide to form a complex consisting of one iron atom surrounded by six cyanides. This complex, known as *ferrocyanide*, is quite stable, and importantly, it can precipitate out of a solution under various conditions. For example, the closely related material known as Prussian blue is a highly insoluble iron cyanide complex that could easily form and precipitate. In addition, salts of ferrocyanide can also precipitate, for example when solutions become concentrated by evaporation. Detailed calculations by planetary scientists Jonathan Toner and David Catling of the University of Washington suggest that ferrocyanide salts could accumulate and precipitate in alkaline carbonate lakes, similar to a class of lakes found on the modern Earth, such as Mono Lake in California and Last Chance Lake in Canada. The source of the dissolved iron is also no longer a mystery. In volcanic regions or around meteorite impact craters, hot water circulates through the fractured rock of Earth's crust, leaching metal ions, including iron, from the rock. As this groundwater is heated, for example by underlying hot magma, it rises to the surface, bringing its cargo of dissolved metals to lakes and ponds. There, iron from Earth's crust can link up with cyanide from Earth's atmosphere, forming ferrocyanide complexes, which can then accumulate over thousands of years. Above a certain concentration, ferrocyanide salts begin to precipitate out, or alternatively, shallow lakes often dry up—both processes that leave behind a bed of sediments mixed with ferrocyanide salts. In this way, huge reservoirs of iron-complexed cyanide could accumulate and be stored for long periods of time, in the geothermally active regions that would have been quite common on the early Earth's surface.

This still leaves us with the question of how a thick bed of dried-up ferrocyanide mixed with mud can be turned into the concentrated stew of the reactive chemicals we need to begin the synthesis of the building blocks of biology. Once again, there is a straightforward answer that lies in one of the most routine geological processes—the transformation of materials under high pressure and temperature. Such a metamorphic transformation turns, for example, soft chalky precipitates of calcium carbonate into that beautiful rock most beloved by sculptors—marble. What kinds of processes could transform our reservoir of unreactive ferrocyanide into the reactive feedstocks we need? Two obvious possibilities are lava flows, which are almost inevitable in volcanically active areas, and asteroid impacts, again almost inescapable on the early Earth. Lava flowing over ferrocyanide-rich sediments would steam, bake, and roast those sediments, releasing the valuable cyanide from the iron grip of its bound form, and in turn transforming some of that cyanide into related reactive species such as cyanamide. Similarly, the impact of a moderately sized asteroid would also provide the heat and pressure necessary for the same chemical transformations. As an amusing aside, most of the scientific literature on the thermal transformations of ferrocyanide salts was published over a hundred years ago! It's just that the potential relevance of ferrocyanide to prebiotic chemistry wasn't recognized until recently. Later in the process, after the hot lava had solidified and cooled, groundwater would begin to slowly percolate through the underlying sediments, carrying away a highly concentrated mixture of cyanide, cyanamide, and other cyanide derivatives.

To summarize the emerging picture, here are the steps we envision. Dilute cyanide first rains from the atmosphere. It is then captured in the form of ferrocyanide by iron brought to the surface via water circulating through fractured rock. The ferrocyanide accumulates in beds of sediments. At some later time, these cyanide-rich sediments are thermally processed by lava flows or meteorite impacts, and a concentrated mixture of reactive feedstock molecules and leftover ferrocyanide is finally released and transported away by circulating groundwater.

## Let There Be Light and Sulfur

To understand what comes next, we have to once again consider the planetary context in which all of this prebiotic chemistry is taking place. As the groundwater bearing its rich load of reactive carbon-nitrogen compounds emerges from the underground darkness of Earth into springs and streams, flowing into shallow lakes and ponds, it will be exposed for the first time to the light of the Sun. The solar radiation impinging on the early Earth would have been a prodigious source of energy, powerful enough to drive a wide range of chemical transformations. Some of these reactions are simple, some complex; some productive, and others destructive. How can we tell whether UV radiation is going to be helpful or harmful? Whether it will drive fruitful synthesis, or simply destroy everything useful? Figuring this out is not a simple task, and a combination of careful experimental measurements with detailed theoretical modeling is required to gain insight. Not surprisingly, this is an ongoing investigation that is engaging many scientists. Still, from the results so far we can distill a few important trends and lessons. Perhaps the most consequential property is the energy of the UV photons. Highly energetic UV radiation (of short wavelength, close to that of X-rays) tends to be the most destructive, because each photon carries enough energy to break up molecules into smaller pieces. In contrast, photons of much less energetic UV light (longer wavelengths, close to that of visible light) don't have sufficient energy to break chemical bonds, so they affect chemistry to a lesser extent. In the middle range is where things get interesting, because these medium-wavelength UV photons can break several but not all chemical bonds. As a result, some compounds will be destroyed, others altered, and still others unaffected, in complex and somewhat unpredictable ways that depend on the precise spectrum of the UV photons, the intensity of the radiation, and the details of the chemical environment. Given all of these caveats, we have to carefully examine how (or even whether) the power of UV radiation could have led to the synthesis of useful compounds on the early Earth.

One of the simplest and most productive consequences of UV irradiation is a very well-studied process, in which a photon of UV light is absorbed by a ferrocyanide complex. As a result, the iron atom of the complex is excited to a higher energy state, and one of the atom's electrons is ejected. This electron initially shoots out at a high speed, but after colliding with a number of water molecules, it slows down and it eventually—in reality after a few nanoseconds (one nanosecond is a billionth of a second)—becomes surrounded by a shell of water molecules. This "aquated electron" is relatively stable, in the sense that it typically lasts for up to several millionths of a second before being absorbed by some other molecule. If the water contains dissolved cyanide, the electron can attach to the cyanide, initiating a series of reactions that ultimately transforms the cyanide into formaldehyde and ammonia. You may have noticed that curiously, for reasons that are not yet entirely clear, prebiotic reactions seem to be dominated by the chemistry of what we identify today as noxious and poisonous substances. The formaldehyde generated by the initial photochemical process reacts very rapidly with a nearby cyanide molecule, producing a simple cyanohydrin. This molecule was for decades believed to be a useless dead-end product because it is relatively stable and unreactive. As a result, it was thought that its formation should be avoided at all costs, in order to steer clear of consuming valuable cyanide and converting it into unproductive junk.

In another conceptual advance, the Sutherland group showed that, just like cyanide, the CN group of this cyanohydrin molecule could also absorb an aquated electron. It would then go through an analogous series of reactions, this time being converted into the simplest sugar, the two-carbon glycolaldehyde that we met before, when trying to produce 2AO. Overall then, the simplest one-carbon aldehyde (aldehydes are organic compounds in which carbon shares a double bond with oxygen)—the famously toxic formaldehyde—has been transformed into a two-carbon sugar. This simple sugar, in itself, turns out to be a useful building block for the assembly of more complex molecules, including larger sugars. Indeed, the same series of reactions that transformed formaldehyde into a two-carbon sugar can be repeated, and by reaction with another cyanide molecule, the two-carbon sugar is transformed into the three-carbon sugar glyceraldehyde, which helped us turn 2AO into RAO earlier.

But let's not be distracted by the complicated names of chemical compounds. The reactions we have described constitute a really striking finding, for several reasons. First, this three-carbon sugar is a central metabolite in modern biology, being one of the key molecules on the pathway by which glucose is metabolized into smaller fragments, thereby providing the energy needed to drive cellular processes. Second, glyceraldehyde (with three carbons) and its smaller precursor, glycolaldehyde (with two carbons), are the starting materials we need to begin the assembly of nucleotides, with their five-carbon sugar—ribose. Interestingly, this process for converting cyanide into the simple sugars can be made even more efficient by invoking a kind of "recycling" chemistry. When a ferrocyanide complex is excited by UV light and emits an electron, it is converted to an oxidized form called ferricyanide. Unless this ferricyanide can be converted back to ferrocyanide, the generation of aquated electrons will grind to a halt when the ferrocyanide is used up. However, the sulfur-containing gas $SO_2$ (sulfur dioxide) dissolves in water to form sulfite and bisulfite, which can recycle ferricyanide back into ferrocyanide. Where does this $SO_2$ come from? This question brings us back once again to the importance of thinking of the geological setting in which these reactions are taking place. If we consider that a likely environment was volcanically active, we recognize immediately that volcanic outgassing can lead to the release of huge quantities of sulfur dioxide. The release of gases that are dissolved in molten rock at high pressure can be quite dramatic as the magma approaches the surface and the pressure decreases. Recall that the explosive Pinatubo volcanic eruption in 1991 released so much $SO_2$ that the resulting atmospheric haze measurably cooled Earth for two years. Sulfur chemistry turns out to be critical for the efficient operation of the prebiotic conversion of cyanide into simple sugars. The entire process is now referred to as *cyanosulfidic photoredox chemistry*, to emphasize the combined synergistic importance of cyanide, sulfur, and UV light.

## Constructing Nucleotides

A closer examination reveals, however, a multitude of problems that have to be addressed before we can conclude that we have a realistic route for making nucleotides. To demonstrate the difficulties, we'll consider two of these problems here. First, those two- and three-carbon sugars are reactive, and glyceraldehyde poses a particular problem since it undergoes a spontaneous rearrangement (a so-called isomerization reaction) from an *aldehyde* (where the C=O group is at one end of the carbon chain) to a *ketone* (where the C=O is in the middle of the molecule). The disturbing result of this rearrangement is that, left to its own devices, more than 99.9 percent of the glyceraldehyde in solution will isomerize to this ketone product, which will not take part in the synthesis of our critical crystalline intermediate, RAO. Researchers therefore had to discover whether the desired two- and three-carbon sugars could be stabilized, so that they could accumulate to high concentrations without forming undesired side products. Fortunately, it turned out that there are at least two ways in which this can be achieved. The first is incredibly simple. Atmospheric $SO_2$ dissolves in water, especially mildly alkaline water, to form bisulfite, which comes to the rescue by reacting with aldehydes (including our simple sugars) to form stable complexes. Moreover, this reaction is reversible, meaning that bisulfite addition products can accumulate, but the free sugar can be slowly released, thus allowing desired reactions to proceed.

Interestingly, there is another, albeit more complicated, way of stabilizing sugars so that they can accumulate in a stable reservoir. This approach, discovered in the laboratory of chemist Matthew Powner at University College London, also involves sulfur chemistry. In this case the key molecule is the sulfur analog of 2AO (the precursor of nucleotides), which we can refer to as 2AT. It turns out that 2AT has the property of reacting with our simple sugars to make stable complexes, which are beautifully crystalline. In a mixture of sugars, 2AT reacts most rapidly with the two-carbon sugar glycolaldehyde, so that a layer of crystals of the 2AT adduct precipitates out of solution. Then, more slowly, 2AT reacts with glyceraldehyde, giving a second layer of crystals. 2AT can even rescue the undesired ketone isomer of glyceraldehyde, because that isomer slowly converts back to glyceraldehyde, which is then removed from the solution as its crystalline 2AT adduct. Which of these two ways of stabilizing sugars is the "real" way that might have happened on the early Earth? Both have certain advantages—the bisulfite process is simpler, but the 2AT process gives us the added potential benefit of a way to physically separate reservoirs of the two- and three-carbon sugars. Whether such a separation is necessary remains to be seen.

Now that we have found a way to keep both our two- and three-carbon sugars in a stabilized form, either in solution or as crystalline reservoirs, the question is what could have happened next. As we have discussed above, glycolaldehyde reacts with the nitrogen-rich compound cyanamide, to generate the reasonably stable intermediate we called 2AO for short. 2AO can react either with another molecule of glycolaldehyde or with a molecule of glyceraldehyde. These reactions generate a bit of a mess, because there are two products of the reaction with a second glycolaldehyde, and four products of the reaction with glyceraldehyde—only one of which is our favorite intermediate, RAO. Out of these six products, it is RAO that crystallizes out of solution, providing the opportunity for the other five by-products to be washed away while a reservoir of pure RAO accumulates. As an aside, the story around the crystallization of RAO is interesting for another reason, because there is a twist (almost literally!) to how these crystals form on a magnetic surface, which could even provide an explanation as to why only one of the two (left or right) mirror-image forms of the nucleotides were made for life as we know it.

Our main concern now is to consider how a nice, clean reservoir of crystalline RAO could have been transformed into what life really needed, which were the nucleotide building blocks of RNA. As we saw earlier, RAO reacts efficiently with the highly reactive compound cyanoacetylene, which is basically just a molecule of acetylene linked to a molecule of cyanide. The product of this reaction is indeed a nucleoside—a precursor of C that is tantalizingly close to the C that is present in RNA, but is not quite right. We'll come back to that particular obstacle in a moment, but the first question is where would cyanoacetylene come from? The answer is that cyanoacetylene can be formed in a reducing atmosphere that is rich in methane, hydrogen, and ammonia, along with cyanide. (Indeed, cyanoacetylene is also abundant in the atmosphere of Saturn's moon Titan.) However, the very reactivity of cyanoacetylene always made its reaction with RAO seem a little dubious, because it was never clear how enough cyanoacetylene could accumulate in the right place at the right time to drive nucleotide synthesis. Recently, researchers in Sutherland's lab found a solution to this problem as well, in an unexpected place—the chemistry of cyanide itself. Concentrated cyanide, dissolved in its hydrolysis product formamide, and gently warmed, produces a remarkably high yield of adenine (abbreviated as A). As we have noted earlier, adenine, one of the canonical building blocks of RNA and DNA, is simply five cyanides hooked together in the right way. But adenine is not the only product of this synthesis. The other dominant product is a molecule that consists of four cyanides joined together, to produce a flat ring-shaped core with two projecting cyanide groups (-CN), referred to as DCI for short. All of this is a roundabout way of bringing us to the solution to the cyanoacetylene problem. To their surprise, the researchers in Sutherland's group found that DCI reacts rapidly with cyanoacetylene to form a stable adduct that, for the sake of simplicity, we will refer to as CV-DCI. This is an interesting compound that crystallizes out of the reaction mixture in which it is formed as beautiful, flat crystals—once again, a nice stable reservoir of a key reactive compound. Critically, however, CV-DCI is not so stable that its precious cargo of cyanoacetylene is locked away in an unreactive form. Rather, CV-DCI can slowly release cyanoacetylene into the solution, where it can go on to react with RAO to form the "not quite right" precursor to C.

At this point, several difficult problems appear to have been solved, but we're not quite out of the woods yet, because we still don't have C itself, but rather a dehydrated form called an anhydro-nucleoside. Worse yet, this anhydro-C has the C nucleobase pointing below the sugar ring (the so-called α-anomer) instead of being above the sugar as in the β-anomer that is universally seen in biology, so that reaction with water generates the α-anomer of C. This was exactly the problem that stalled anhydro-α-riboC $\xrightarrow{HS^-}$ 2-thio-α-riboC $\xrightarrow{UV}$ 2-thio-β-riboC

progress some sixty years ago. It was at that time that chemist Leslie Orgel and colleagues found that UV light could convert a very small fraction of the α form to the desired β form. This result was frustrating as much as it was exciting, because there did not seem to be any way for RAO to be rescued from the dead-end transformation into α-riboC.

The solution to the problem of how to use the propensity of RAO to self-purify by crystallization came, once again, from thinking about the geological context in which prebiotic chemistry took place.

We have already discussed potential roles of one sulfur-containing volcanic gas, $SO_2$, but there is another sulfur-containing gas that is hard to miss in any volcanically active area, and that is hydrogen sulfide, or $H_2S$. This is the gas that gives the odor to rotten eggs, and it is also one of the most recognizable and dangerous gases released from volcanic lava or subterranean magma. Indeed, one of us (Szostak) was once on vacation on the Caribbean island of Dominica, and tried to visit the famous "boiling lake" of its active volcano. However, the pungent odor of $H_2S$ became overpowering long before the lake came into view, and he was forced to turn back. Believe it or not, the second author (Livio) had an identical experience on Vulcano Island near Sicily. Like its more oxidized cousin $SO_2$, $H_2S$ can also dissolve in water, especially if the water is somewhat alkaline. Thus, water that circulates in the crust and comes into contact with gases released from magma can accumulate hydrogen sulfide, which can do many things of interest. For example, hydrogen sulfide reacts with metal ions such as ferrous iron and precipitates them as the corresponding metal sulfide, which in the case of iron is the gold-like iron pyrite. However, if some of the sulfide remains in surface waters, even more interesting reactions can occur, which brings us back to our RAO conundrum. As noted above, RAO reacts with cyanoacetylene (coming from its slow-release reservoir of CV-DCI) to form the almost but not quite right "anhydro" form of C, where the problem is that the ring of C is pointing down from the sugar ring, instead of up as in the biological form of C. When this anhydro-C is hydrolyzed by water, we end up with the so-called α-anomer of C, and we have no good way to convert that to the desired β form by flipping the ring from below the sugar to above it. Well, researchers discovered that sulfide can also attack this anhydro precursor, generating a product in which one of the oxygen atoms of the nucleobase is replaced with a sulfur atom. This might seem like a step backward, since now we have two problems: the ring is still in the wrong position, and in addition it is now modified by the presence of a sulfur atom. Almost incredibly though, it is this sulfur atom that comes to the rescue, because now mild UV irradiation excites the molecule and efficiently flips the ring to the desired upward position. Continued UV exposure under mildly alkaline conditions leads to the loss of the sulfur from the product, leaving behind the exact natural form of the ribonucleoside C. Moreover, that same exposure to additional UV radiation in an alkaline environment also converts C to U, giving us two of the four canonical building blocks of RNA!

To recap briefly, we have seen how cyanide can be converted into the simple two- and three-carbon sugars, which can then be stabilized by reacting with either bisulfite (from volcanic $SO_2$) or 2AT. Cyanamide then reacts with these sugars to form a complex mix of products, of which one isomer—RAO—spontaneously crystallizes from solution, thereby accumulating as a reservoir of purified material. The RAO then reacts first with cyanoacetylene (derived from another crystalline reservoir, CV-DCI), and then with hydrogen sulfide, to form a

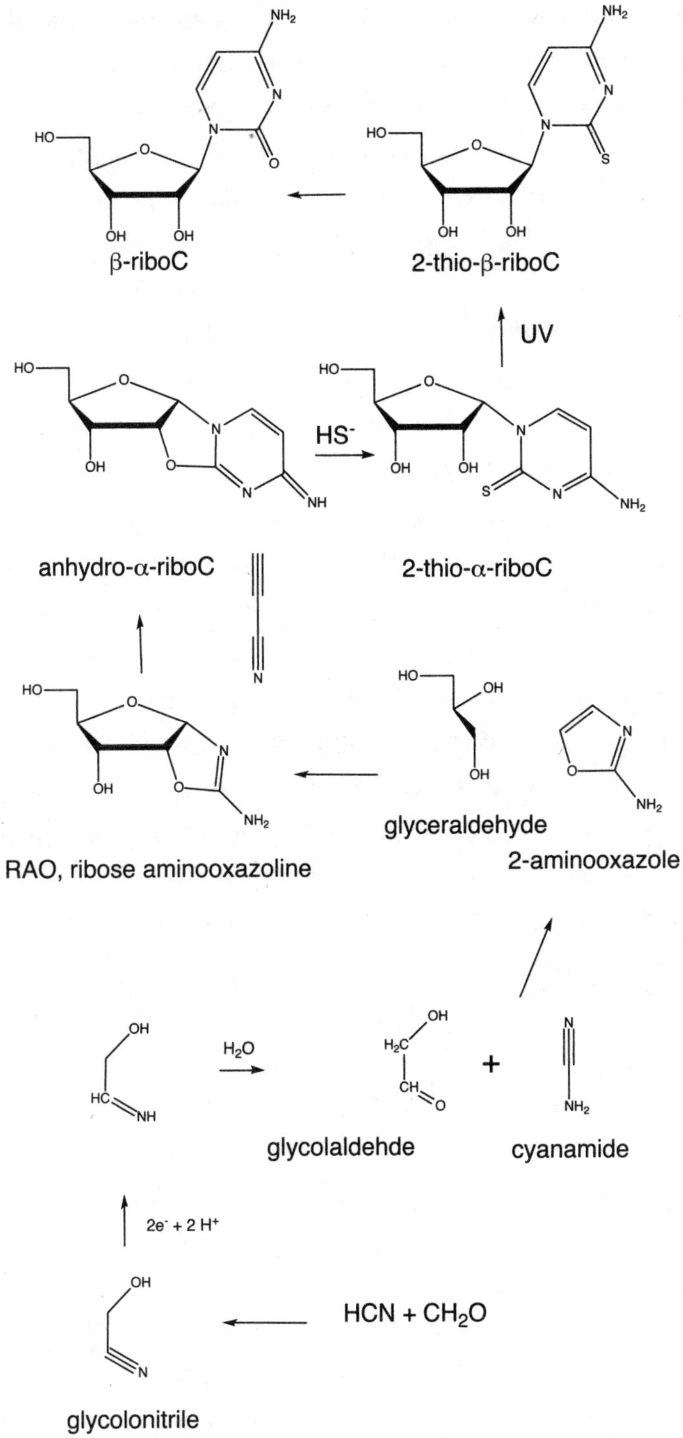

$NH_2$ N HO N O O OH OH

β-riboC

$NH_2$ N HO N O S OH OH

2-thio-β-riboC

UV

HO O N OH O N NH

HO O N OH OH S N $NH_2$

$HS^-$

anhydro-α-riboC

2-thio-α-riboC

N

HO O N OH O $NH_2$

RAO, ribose aminooxazoline

HO OH OH

glyceraldehyde

N O $NH_2$

2-aminooxazole

OH HC NH

$H_2O$

OH $H_2C$ CH O

glycolaldehde

+

N $NH_2$

cyanamide

$2e^- + 2\,H^+$

OH N

glycolonitrile

$HCN + CH_2O$

sulfur-containing nucleoside, which is converted to the correct structure (anomer) by exposure to UV light. Finally, continued UV exposure in alkaline water generates a mixture of the biologically relevant C and U nucleosides. The overall pathway is summarized in the figure on p. 63.

*Perhaps the most important aspect of this series of experimentally demonstrated reactions is that cutting-edge research has replaced the old-fashioned concept of a "prebiotic soup" with a series of steps in which intermediates are stabilized while in solution, and purified by crystallization. These crystalline intermediates, which are essentially organic minerals, can accumulate over time, until they are either destroyed or go on to react in the next step of the pathway.*

An obvious question arises: What are the chances that all of these steps can occur in nature in the right order, under the right conditions, to generate the final biological building blocks? This question is very difficult (if not impossible) to answer from the chemistry alone. Indeed, it is possible that the hardest—that is, slowest—aspect of the origin of life is having a series of compounds made and accumulated as reservoirs, which are then brought together in the right order so that the biologically relevant products are generated. It is in the attempts to answer this question that the astronomical searches for extraterrestrial life may provide some insights, if, for example, they show that life in our galaxy is not rare.

Making two of the four RNA nucleosides is not all that is needed to construct a living protocell, not by a long shot. The very success in making C and U (the *pyrimidine nucleosides*) begs the question, what about the remaining *purine nucleosides* A and G? Many new ideas are being explored, but no clear answer is yet in sight. Another critical issue is how to attach a phosphate to the nucleosides, to generate the nucleotides that are the real subunits of an RNA chain. This is another area that seems ripe for advance. The problem here is how to add phosphates to a nucleoside (this is called phosphorylation) in the right place, since there are three hydroxyl (-OH) groups on the ribose sugar, any of which could in principle become phosphorylated. Most approaches to the required phosphorylation are quite harsh and nonspecific, meaning that phosphates are added at any or all of those three positions, generating a mixture of compounds. In biology, the nucleotide building blocks of RNA and DNA always have their phosphate on a particular -OH (sticking out above and away from the rest of the sugar), but frustratingly, it seems chemically easier to get phosphates onto the other two hydroxyls, in which case, the phosphate tends to cyclize—generating what is known as a cyclic phosphate. Intriguingly, this is the same product that is generated from the hydrolysis of RNA. One possibility could be that nucleotides or short strands of nucleotides with a terminal cyclic phosphate are the actual primordial building blocks of RNA, and that the particular phosphorylation we see in biology today is a later "invention" of evolution. This idea is consistent with the fact that such short strands can assemble on a template (when one strand serves as a mold for the other) to generate longer products.

On the other hand, the slow rate and poor yield of such assembly reactions could equally well be interpreted as evidence that the process cannot be relevant to RNA synthesis. It is certainly the case that template copying by nucleotides (or short strands) with an activated phosphate at the right location is much faster and more efficient. An early attempt to direct phosphorylation to the right position invoked the presence of borate (a salt of boric acid), which complexes with the other two hydroxyls and prevents them from becoming phosphorylated. This works reasonably well, but the plausibility of having enough borate around in the early Earth environments where nucleotides were being made is highly debated. Whether or not some other way will be found of making the phosphorylation reaction both gentler and more specific is currently unknown, but the search for such a path is clearly a priority in the field.

As we have seen, producing the building blocks of RNA in the laboratory encountered many challenges. Nevertheless, careful, creative thinking, coupled with extensive empirical work, managed to overcome many (although not all) of the obstacles. But this is just the beginning. The experimental production of many other components of living cells needs to be achieved before we can claim that we understand the emergence of life.

APPENDIX

# Chemical Structures and Reactions

This appendix consists of a series of chemical diagrams that illustrate how the building blocks of RNA can be assembled from simpler starting materials, as described in the text.

## How to "Read" Chemical Diagrams

Chemical structures are conventionally drawn in a kind of shorthand format that can be confusing at first. However, the rules of this shorthand are quite simple. Each line represents a bond between adjacent atoms. A single line represents a single bond, a double line represents a double bond, and a triple line represents a triple bond, so that much is fairly straightforward. A little less obvious is the rule that carbon atoms are seldom explicitly represented as a C; instead, any place where two bonds meet is a carbon atom. As for hydrogen atoms, we don't even bother to write them down, but it's OK, you can figure out where the hydrogens are, because carbon atoms always form four bonds. As an example, the vertex between a single line and a double line represents a carbon atom with the three bonds shown, and the missing fourth bond is to an implied hydrogen atom. Sometimes this can be a little extra tricky, as in the structure of cyanoacetylene in the first figure below. Cyanoacetylene is drawn as a linear molecule, because it really is a linear molecule. You will see two triple bonds: at the top of the molecule, the "cyano" piece is a carbon atom joined to a nitrogen atom by a triple bond, while at the bottom of the molecule, the acetylene piece is two carbon atoms joined by a triple bond. These two pieces are joined together by a single bond. Thus, the two carbon atoms in the middle of the molecule are not explicitly drawn, but are implied by the junction of the triple and single bonds.

## 1. A synthesis of the nucleobase C, a.k.a. cytosine

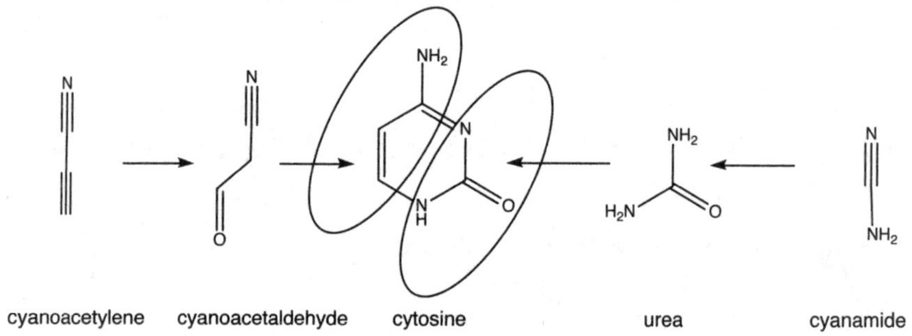

cyanoacetylene cyanoacetaldehyde cytosine urea cyanamide

Two feedstock molecules, cyanoacetylene on the left and cyanamide on the right, are composed of carbon, nitrogen, and hydrogen. Both react with water to form hydrated derivatives: cyanoacetaldehyde on the left and urea on the right. These two compounds can join together to form cytosine, the nucleobase component of the cytidine nucleotide. The upper-left oval encloses the atoms that come from cyanoacetylene, while the lower-right oval encloses the atoms that come from urea.

## 2. Synthesis of RAO

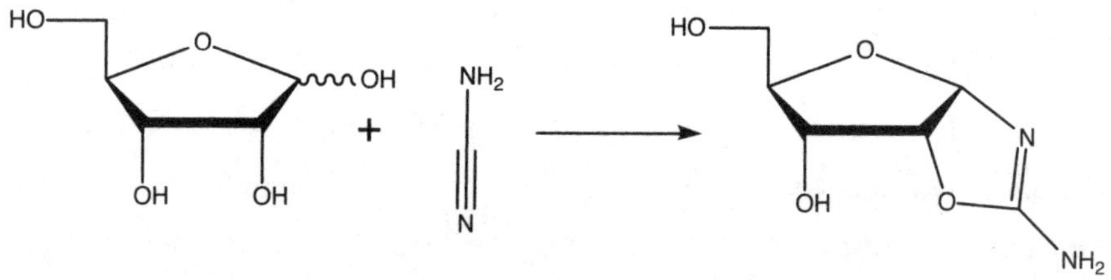

The sugar ribose reacts with the feedstock cyanamide to form the key intermediate in nucleotide synthesis, RAO, a.k.a. ribose amino-oxazoline.

## 3. Synthesis of α-cytidine

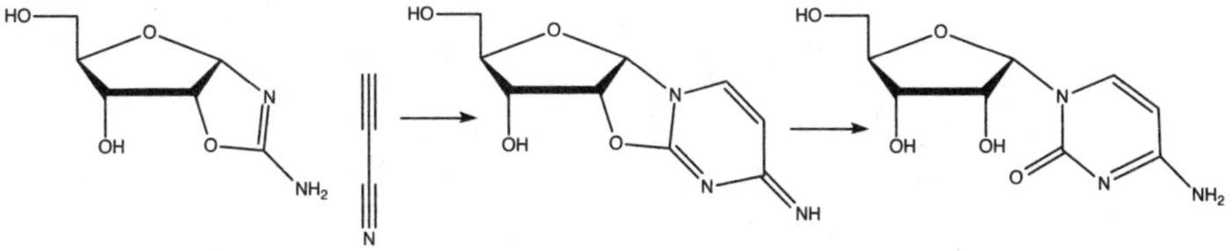

RAO (left) reacts with cyanoacetylene (to the right of RAO) to give α-anhydrocytidine (center), which hydrolyzes in water to yield the α-anomer of cytidine (right).

## 4. Synthesis of 2-aminooxazole

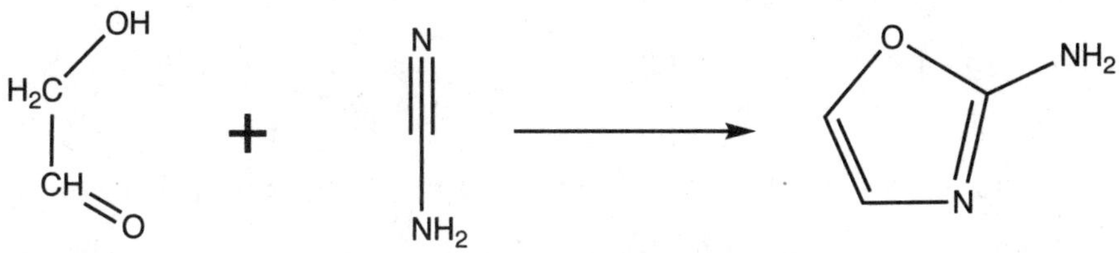

Glycolaldehyde + cyanamide combine to yield 2-aminooxazole.

## 5. A different way to make RAO

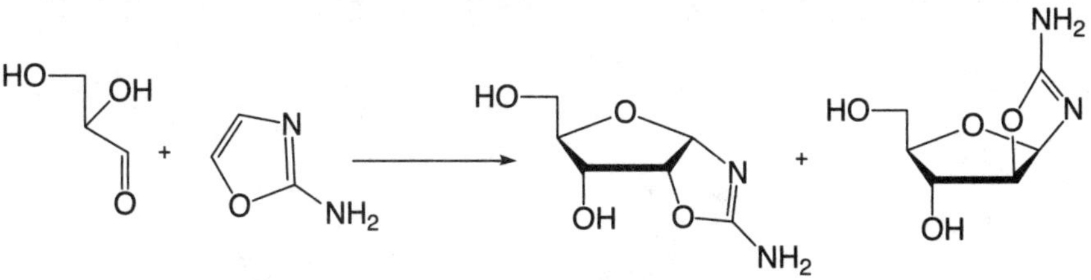

Glyceraldehyde + 2AO combine to form a mix of RAO and AAO, plus other isomers that are not shown.

## 6. A way to make riboC from araC

Anhydro-araC with a phosphate on the 3'-OH transforms to riboC with a 2'-3'-cyclic phosphate.

## 7. Synthesis of glycolonitrile

$$H_2C{=}O + HC{\equiv}N \longrightarrow HOCH_2C{\equiv}N$$

*formaldehyde + cyanide* → *glycolonitrile*

Two reactive feedstock molecules, formaldehyde and cyanide, react with each other to form glycolonitrile. For many years this relatively unreactive molecule was thought to be a "dead-end" product to be avoided at all costs.

## 8. Transformation of glycolonitrile into the simplest sugar, glycolaldehyde

$$HOCH_2C{\equiv}N \xrightarrow{2e^- + 2H^+} HOCH_2CH{=}NH \xrightarrow{H_2O} HOCH_2CH{=}O$$

*glycolonitrile* → *imine intermediate* → *glycolaldehyde*

## 9. Isomerization of the sugar glyceraldehyde into dihydroxyacetone

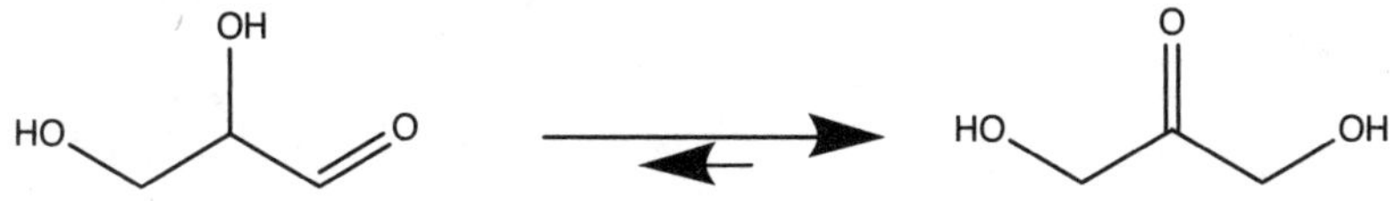

Glyceraldehyde Dihydroxyacetone

## 10. 2AO and its close relatives 2AT and 2AI

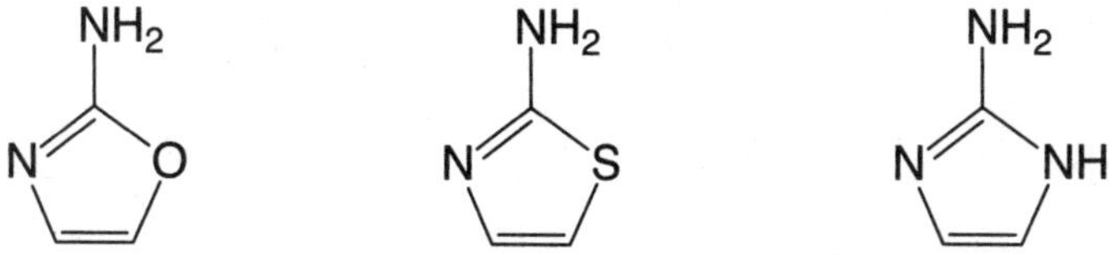

2AO 2AT 2AI

2AO is 2-aminooxazole. An azole is a five-atom ring containing a nitrogen atom (N is *aza*). An oxazole is a five-atom ring containing an oxygen and a nitrogen atom, hence oxazole.

2AT is 2-aminothiazole. A thiazole is a five-atom ring containing a sulfur (*thia*) and a nitrogen atom, hence thiazole.

## 11. Synthesis of CV-DCI

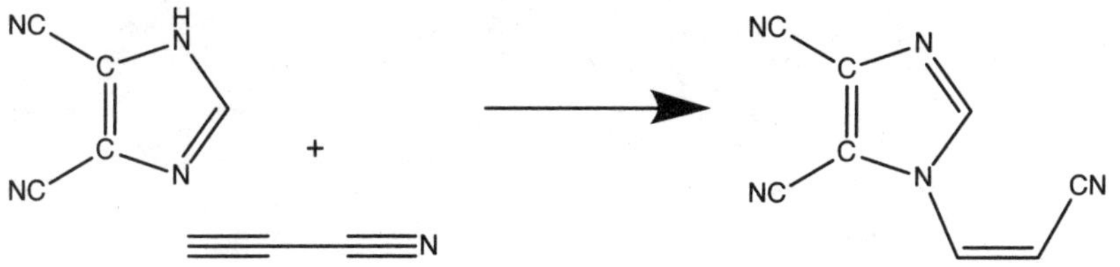

Dicyanoimidazole and cyanoacetylene react to form the adduct CV-DCI, a.k.a. cyanovinyl-dicyanoimidazole. CV-DCI crystallizes as flat plates, which can be viewed as a stable storage form or reservoir of cyanoacetylene.

CHAPTER 4

# The Origin of Life

## Amino Acids and Peptides

*It is one of the more striking generalizations of biochemistry—which surprisingly is hardly ever mentioned in the biochemical textbooks—that the twenty amino acids and the four bases, are, with minor reservations, the same throughout Nature.*

—Francis Crick, Nobel Lecture

Any attempt to discover the origin of life on Earth has to identify a chemical pathway leading to the production of proteins. Proteins with enzymatic activities catalyze the vast array of metabolic reactions required to synthesize the materials for new cells. In addition, proteins that assemble into fibers control cell shape and such dynamic processes as cell movement and cell division. All of these proteins are made in the complex process of *translation* in which the cellular machine known as the ribosome translates the genetic information in messenger RNAs, which are long strings of nucleotides, into proteins, which are ordered strings of amino acids. This task of translation is mediated by the genetic code that relates sequences of RNA codons to sequences of amino acids. The process is far too complicated to have been fully operational during the origin of life, but the earliest seeds of the procedure must be reflected in the chemistry of amino acid synthesis, the chemical reactions that lead to the production of peptides (short chains of amino acids), and ultimately in the marvelous chemistry that links amino acids with RNAs. Let's then go back to the beginning, and consider the chemical routes that could lead to amino acids.

We'll start by examining a famous breakthrough experiment, and by mining its details for the insights that it can provide into realistic pathways for amino acid synthesis on the early Earth. In 1952, chemist (then graduate student) Stanley Miller, under the supervision of Nobel Prize laureate Harold Urey, designed an experiment at the University of Chicago, which was supposed to investigate the conditions thought to have existed on the early Earth. The ensuing discovery, that an electrical spark discharge in an artificial "atmosphere" of hydrogen, methane, ammonia, and water led to the synthesis of two amino acids, was considered revolutionary at the time. The realization that compounds as central to life as amino acids—the quintessential building blocks of proteins—could be produced in such a simple way shocked the chemistry world, and inspired decades of subsequent research. Urey, by the way, graciously gave Miller the full credit for the experiment. The Miller-Urey result is commonly described as demonstrating amino acid synthesis, but that is, in fact, not quite correct. The actual products generated in the reaction were the so-called α-amino nitriles, which are closely related to the α-amino acids (except that what is known as a nitrile group is attached to the central α-carbon in place of the acidic carboxyl group). This in itself is not really a problem, since nitriles slowly hydrolyze in water to carboxylates (compounds with a carboxyl group). This hydrolysis reaction is greatly accelerated by strong acids, which is what Miller used to unmask the amino acids from their nitrile

$$H_2C{=}O \; + HC{\equiv}N \; + NH_3 \longrightarrow H_2N\text{-}CH_2\text{-}C{\equiv}N$$

*formaldehyde + cyanide + ammonia* *glycine nitrile*

precursors. In fact, sulfide, common in volcanic regions, can also accelerate the hydrolysis of nitriles to carboxylic acids, by going through a more reactive intermediate. This means that the conversion of α-amino nitriles to amino acids is something that should be expected to occur on the early Earth, either slowly or quickly, depending on the specific chemical environment.

The upshot of all of this is that if we want to understand how amino acids can be made, what we really need to understand is how α-amino nitriles can be made, and the simplest way of doing just that was discovered in the mid-nineteenth century by the German chemist Adolph Strecker. He showed that when aldehydes (molecules with a C=O group) are mixed with cyanide and ammonia, α-amino nitriles are generated. This reaction, commonly referred to as the Strecker reaction, is well known and widely used. The important point to notice here is that we immediately recognize a link between the synthesis of the precursors of amino acids and the precursors of the nucleotides, since both involve reactions of aldehydes with cyanide.

As we have previously seen, when cyanide attacks aldehydes in water, the product is a cyanohydrin. Subsequently, reduction of the -CN group (for example by aquated electrons) to an aldehyde generates a sugar, and the simple sugars can go on to assemble nucleotides. Moreover, in a slight variation on this theme, when cyanide attacks an aldehyde in the presence of ammonia, an α-amino nitrile is generated, and in this case hydrolysis of the nitrile generates an amino acid. Specifically, formaldehyde, cyanide, and ammonia combine to give glycine nitrile, the α-amino nitrile corresponding to the amino acid *glycine*. This amazing connection between the synthesis of nucleotides and that of amino acids is a major clue that the building blocks of biology *glycolaldehyde* $\xrightarrow{CN^-, NH_3}$ *serine nitrile* $\xrightarrow{HS^-, H_2O}$ *serine*

$\downarrow$ $2e^- + 2H^+$

*acetaldehyde* $\xrightarrow{CN^-, NH_3}$ *alanine nitrile* $\xrightarrow{HS^-, H_2O}$ *alanine*

could all have been synthesized at the same time in similar and possibly nearby environments!

Having traced back the chemical origins of the amino acids through α-amino nitriles to aldehydes, we can now reformulate the search for prebiotic pathways to the amino acids to a search for the corresponding aldehydes. For some of the simplest amino acids, this is pretty straightforward. For example, the easiest to produce (and therefore likely most abundant) aldehyde would be the one-carbon formaldehyde. As we have seen earlier, formaldehyde reacts with cyanide to make the simplest cyanohydrin, glycolonitrile. In the presence of ammonia, glycine nitrile would be made, which in turn would hydrolyze to make the simplest possible amino acid, namely glycine.

As an interesting etymological note, the *gly-* prefix of this set of compounds comes from "glyco," or sugar, and echoes the sweetish taste of glycine. As the simplest amino acid, we would expect glycine to be the most abundant amino acid in prebiotic environments. Still, to make more interesting peptides, we would need more than just glycine. It's easy to see the origins of the next two amino acids, *serine* and *alanine*. Serine derives directly from the two-carbon sugar glycolaldehyde we encountered in Chapter 3, through the same Strecker synthesis—that is, a reaction with cyanide and ammonia, followed by hydrolysis of the nitrile to the carboxylic acid group. Making alanine requires one more reaction—the reduction of glycolaldehyde to acetaldehyde. This reduction involves the replacement of the hydroxyl group (composed of an oxygen atom bonded to a hydrogen atom) of glycolaldehyde with a hydrogen atom. Interestingly, this reduction reaction is driven by exactly the same chemistry as the reduction of a nitrile to an aldehyde. In both cases, the aquated electrons obtained by the UV irradiation of ferrocyanide (as explained in Chapter 3) are acting as powerful reducing agents to drive analogous reactions. Once acetaldehyde is generated in this way, the same Strecker synthesis will yield the common amino acid alanine.

So far we have seen how the three simplest amino acids can be derived from the same compounds that go on to generate the nucleotides. Let us now consider the next two, slightly larger and more complex amino acids, *threonine* and *valine.* To get to these and other even more intricate amino acids, we will invoke the repeated application of the same small sets of reactions. While the accumulation of hard-to-pronounce compound names may look intimidating, you'll notice that the principles involved are very simple. The first reactions include the attack of cyanide on an aldehyde in water to generate the larger cyanohydrin product or, in the presence of ammonia, to generate an α-amino nitrile. The second set consists of the reduction reactions driven by aquated electrons, including reduction of a -CN nitrile to an aldehyde (via an intermediate) and reduction of an -OH group to -H. The third set involves the attack of hydrosulfide (HS-) on a nitrile group to give a group that can then be either hydrolyzed to a carboxylate or reduced to an aldehyde. Using these three very basic types of chemical reactions, we can easily generate both *threonine* and *valine.* To get threonine, we begin with the same acetaldehyde used to make alanine via the Strecker synthesis, except that this time we will allow the acetaldehyde to react with cyanide to generate a new cyanohydrin, then reduce its nitrile to form a new aldehyde. We can then repeat the cycle of reaction with cyanide, and if ammonia is present, a new α-amino nitrile is formed that is the direct precursor of threonine.

Valine can be formed in a very similar manner, but with an interesting twist. To form valine, we begin with glyceraldehyde, the three-carbon sugar that is a key ingredient in nucleotide synthesis. Recall that glyceraldehyde is unstable, and spontaneously isomerizes into a ketone where the carbon with a double bond to an oxygen atom is in the middle of the molecule and not at its end. In order to use glyceraldehyde to make nucleotides, we had to prevent that isomerization reaction from happening, or reverse it if it did happen. However, to make valine, that isomerization step is essential. That isomer has two hydroxyl groups, one on each of the carbon atoms at the ends of the three-carbon chain. Those hydroxyls can be replaced with hydrogen atoms, to generate *acetone* (a compound commonly found as a solvent in nail-polish removers). Acetone can be used to generate valine by an initial reaction with cyanide to form a new cyanohydrin, followed by reduction of the hydroxyl to hydrogen and reduction of the nitrile to yet another aldehyde, which is the direct precursor of valine through the Strecker synthesis.

As we consider the synthesis of progressively larger and more complex amino acids, we see more and more of the connections between nucleotides and amino acids. An especially striking example of such a connection is found in the routes to the amino acids *asparagine*, *aspartic acid*, *glutamine*, and *glutamic acid*.

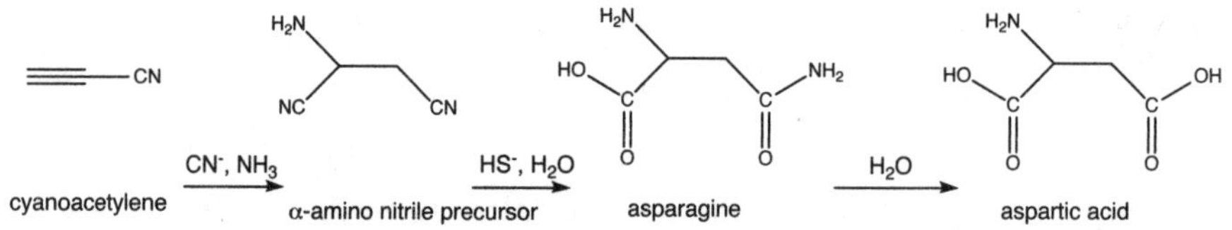

cyanoacetylene → ($CN^-$, $NH_3$) → α-amino nitrile precursor → ($HS^-$, $H_2O$) → asparagine → ($H_2O$) → aspartic acid

All four of these amino acids can be generated by starting with cyanoacetylene, precisely the same reactive starting material that is an essential ingredient in the synthesis of the nucleobases of the C and U nucleosides. In that case, cyanoacetylene, slowly released from its crystalline reservoir form CV-DCI, reacted with RAO to generate an anhydro form of the C nucleoside. However, in an environment rich in cyanide and ammonia, cyanoacetylene can form the α-amino nitrile precursor of asparagine and aspartic acid. This precursor contains two nitrile groups, one next to the amino group, which hydrolyzes to the carboxylate, while the other nitrile is at the opposite end of the molecule. If that second nitrile undergoes partial hydrolysis, asparagine is formed; further hydrolysis to a second carboxylate generates aspartic acid. Moreover, the carbon backbone can be extended by the reduction of one nitrile to an aldehyde followed by the usual addition of cyanide and further reaction to form the one-carbon longer precursor to glutamine, which, after hydrolysis, yields glutamic acid.

If the synthesis of glutamine and glutamic acid wasn't complicated enough, there are two similar but even longer pathways leading to two additional amino acids: *proline* and *arginine*. These amino acids are both interesting in that they play important and very different roles in peptide and protein structure. Proline is the only biological amino acid that has a ring-shaped structure.

As a result, it creates a bend or kink in the structure of a peptide, and also interrupts canonical structures found in proteins such as the

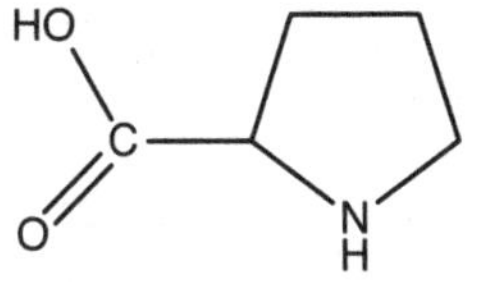

proline

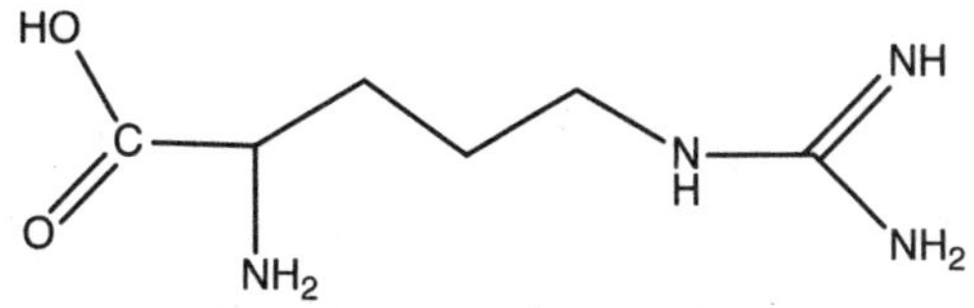

arginine

famous α-helix discovered by Nobel laureate chemist Linus Pauling. Arginine, on the other hand, carries a positive charge on its side chain, and can therefore interact strongly with negatively charged molecules such as RNA. On the face of it, proline and arginine look quite different, and yet they share a common early precursor in this chemical synthesis pathway, and an early divergence from the route leading to the acidic amino acids. While cyanide and acetylene can be coupled in one way to yield cyanoacetylene, they can also be coupled in a different way to yield a compound called *acrylonitrile* (technically, these are oxidative and reductive processes). In general, the elaboration of acrylonitrile into proline and arginine follows the same types of reactions we have been discussing for the other amino acids. However, both pathways also involve some unique steps, such as the reaction to generate the positively charged group of what will eventually become arginine. In contrast, the pathway to proline involves an early generation of the proline ring, followed by a series of more familiar reactions.

So far we have seen that eleven of the canonical twenty protein-forming amino acids found in life on Earth can be generated from the same network of chemical reactions (the one dubbed cyanosulfidic photoredox chemistry) that also generates at least two of the four canonical nucleotide building blocks of RNA. The natural question that arises is what about the remaining nine amino acids found in biology today? A few of these, such as the aromatic amino acids *phenylalanine*, *tyrosine*, and *tryptophan*, are thought to be late additions to the genetic code, in part because there is no obvious prebiotic route to their formation. It is possible that these amino acids emerged during the evolution of cellular metabolism, and were then co-opted for protein synthesis. However, this opinion may change in the future, as new prebiotic routes are discovered. For example, the amino acid *cysteine*, which is relatively unstable, had no known plausible prebiotic synthetic pathway until recently, when it was discovered that cysteine can be derived from an α-amino nitrile precursor of the related amino acid *serine*, in a series of three steps.

To briefly summarize the essence of what we have presented in the last three chapters: We have first seen how a series of conceptual breakthroughs, largely guided by thinking about the geological environments of the early Earth, has led to the discovery of chemical pathways to many (but not yet all) of the key building blocks of biology. Importantly, these pathways give high yields of a restricted set of relevant products, as opposed to earlier approaches that resulted in numerous chemical compounds, with the necessary starting materials for biology present in vanishingly small quantities. However, even these relatively efficient "new" synthetic pathways cannot deliver complicated molecules such as nucleotides in a single prebiotic-soup-type scenario. The old prebiotic-soup concept has now been replaced by the repeated accumulation of precipitated or crystallized intermediates, with short sequences of reactions leading from one intermediate to the next. Crucially, we have shown that both this accumulation of reservoirs of intermediates and the intervening chemical reactions could well have operated on the surface of the early Earth. *The difficult, and therefore slow, step in producing the molecules of life now appears to be not the chemistry per se. Rather, it's the apparently low probability of having precisely the right set of environmental changes, all in the right places and at the right times, so that these reservoirs can accumulate without being destroyed and can then go on to the next chemical transformation, until finally the stage has been set for the emergence of the first living cells.*

# CHAPTER 5

# The Origin of Life

## The Road to the Protocell

*A cell is regarded as the true biological atom.*

—George Henry Lewes, *The Physiology of Common Life*

All biological life on Earth is cellular: large organisms such as us consist of trillions of cells, while simpler organisms such as bacteria or yeast are single cells or small clusters of cells. Cells are in a sense the units of life, since the way life propagates is through the growth and subsequent division of cells. Looking backward in time, therefore, there must have been a first cell, and we can ask ourselves what that cell might have looked like—what its structure and chemical composition could have been. In terms of understanding the origin of life, we would like to know how that first cell assembled from the nonliving chemicals present on the surface of the early Earth, in what type of an environment, and with which kinds of energy sources. Somewhat surprisingly, this self-assembly problem in itself turns out to be rather simple, and the real challenge proves to be understanding how such an incredibly simple cell, with no evolved biochemical machinery, could grow and divide to produce more and more progeny cells. But before delving into these questions, we should ask ourselves why life began with the formation of a simple cell. In other words, we need to clearly identify what it is about cellular organization that is so important.

The essence of cellular structure is that a cell is a compartment—a physically localized set of molecules that are in some way segregated from the rest of the environment. At a very basic level it is easy to see why this is essential. After all, we ourselves are individuals, and we wouldn't want our component parts to simply dissolve and float away, dispersing in the environment. The same holds true for single-celled organisms such as bacteria, and also for the very first cells, where it was important to keep the components of an individual cell from drifting apart. However, there is a subtler and more fundamental reason for the importance of the cell as the unit of life, and that is that *spatial localization is a requirement for the emergence of Darwinian evolution*, and thereby for the evolution of all the diverse forms of life that we see around us. To see why this is the case, let us consider a primordial RNA molecule, endowed by virtue of its sequence with the ability to catalyze some sort of metabolic reaction. For example, let us imagine that this metabolic activity is the synthesis of nucleotides, which could be used to make more RNA. Now, let us further imagine that during the replication of this primordial RNA, an error in copying its sequence occurs, so that the progeny of this RNA molecule are now a mutant version of the original sequence. Such a mutant RNA might now catalyze the synthesis of nucleotides more rapidly than the original ribozyme. If all of this were to happen with free-floating RNAs, the products of this metabolic reaction would simply diffuse into the surroundings, possibly aiding other RNAs that required nucleotides for their replication. In effect, the mutant RNA would not benefit from its superior catalytic ability. Now consider the alternative scenario, in which the original

RNA resides within some sort of localized compartment, while other RNAs reside within their own separate compartments. In this situation, a mutant RNA that happens to acquire a more effective metabolic activity will be able to exploit its superior catalytic activity because the products of its ability, nucleotides in this case, will also remain physically within the same compartment. There they can contribute to the replication of the mutant RNA, but not to other RNAs, thus giving the mutant RNA a fitness advantage.

This argument for the evolutionary advantage of compartmentalization applies to almost any consequence of mutations. As another example, consider an RNA molecule that functions as an RNA polymerase that can help to replicate its own sequence. This is not as fanciful as it may sound because such molecules are being evolved in laboratory experiments. As we noted in Chapter 2, an RNA molecule that can help to replicate its own sequence is known as an RNA replicase. Such RNAs are thought to have been crucial in the early evolution of life. In this example, we need to consider the fact that an RNA molecule must fold up into a specific three-dimensional shape in order to act as a catalyst, just like protein enzymes. On the other hand, in order for an RNA molecule to be replicated, it must unfold, so that the enzyme doing the copying can march along its linear sequence as it synthesizes the copy. That is, for an RNA molecule to copy itself, what we need is two RNA molecules with the same sequence (or with complementary sequences). One of these molecules must be unfolded so that it can act as the template being copied, while the other is folded and acts as the copying enzyme. This requirement immediately highlights the need for compartmentalization—in the absence of some physical co-localization, these two RNAs would drift apart, and the copying process would happen rarely, if at all. Now imagine an RNA molecule that is an RNA polymerase, but is floating freely in a solution where it is surrounded by a multitude of other unrelated RNAs. It would presumably busily make copies of those other RNAs. Worse yet, a mutant version that was a superior RNA polymerase ribozyme would be better at copying other unrelated RNAs, but would not itself benefit from its higher RNA polymerase activity. On the other hand in the cellular scenario, a mutant RNA with enhanced replication ability would indeed benefit from its advantage, because it would be replicating itself or, at least, molecules related to itself by descent from a common ancestor.

There is yet a third rationale for the cellular basis of life, which is experimentally well established: resistance to parasites. Among the earliest examples of experimental molecular evolution are brilliant experiments carried out in the 1960s by University of Illinois, Urbana-Champaign, molecular biologist Sol Spiegelman and his collaborators. These experiments made use of a protein enzyme with the ability to replicate the RNA genome of a bacterial virus with the odd name of Qβ. The Qβ virus maintains its full-length genome throughout multiple generations of growth in bacterial cells, that is, replication within compartments. But what Spiegelman observed, in a series of experiments, was that propagation of the viral RNA genome in solution led to the rapid emergence of much smaller parasitic RNAs. Because these parasitic RNAs are smaller than the full-length genomic RNA, they replicate much faster and quickly take over the population. These shorter RNAs arise during viral replication, when the polymerase enzyme accidentally skips over part of the sequence, generating a deletion mutant. This also happens when the viral RNA is replicating within cells, but the deletion mutants cannot spread throughout the population because they are defective. Although it may seem strange to use viruses as an argument for the role of cellular compartmentalization as a defense against parasites, the lesson is clear: replication of RNAs in solution (i.e., not in compartments) results in collapse of the population as smaller and faster-replicating defective RNAs outcompete larger RNAs. In contrast, defective deletion mutants (parasites of the virus in our example) cannot take over and destroy the entire population.

Having established the necessity of a cellular origin of life, we are now in the position to ask what *kind* of physical compartments would be best suited for the first cells, which we will refer to as protocells. Let us first examine whether we can use the modern-day biological solution to this problem as the basis for a prebiotic answer. We will return to the question of potential alternative types of compartments later on. In the previous chapters, we focused on the discovery of increasingly realistic chemical pathways to the nucleotides, which are needed to generate the genetic molecules of a protocell, and the simple amino acids, which may play multiple roles in protocell function. In addition to nucleotides, RNA, amino acids, and peptides, the other fundamental component of a protocell that is modeled on modern biology is its *membrane* envelope. Biological membranes are composed of a diverse collection of molecules with the common property of being "amphiphilic," meaning that one end is hydrophilic and likes to be exposed to water, while the other end is hydrophobic and likes to be kept away from water. This property results in the spontaneous self-assembly of bilayer membranes, which are composed of two layers of lipids (fatty compounds), in which the hydrophilic sides of both layers face the surrounding water, while the hydrophobic parts form the middle of the membrane and are hidden from water. Modern lipids are generated by metabolic pathways inside cells, but how would membrane-forming molecules have been generated on the early Earth? This is in fact one of the thorniest remaining mysteries in the field of prebiotic chemistry. The simplest class of molecules that can form membranes are the fatty acids, which are essentially a hydrophobic hydrocarbon chain ending in a hydrophilic carboxyl group (a carbon atom with two oxygens attached). Although these molecules are structurally quite simple, understanding how to make them in any plausible early Earth environment remains a challenge. All of the most commonly discussed models for prebiotic fatty acid synthesis—from formation during meteorite impacts to synthesis deep inside Earth—seem inadequate to explain the synthesis of fatty acids at the high concentrations needed for assembly into membranes. Although new pathways are currently being explored, much more work is still required to resolve this issue.

Given that we lack a clear understanding of how even such simple molecules as fatty acids might have been synthesized on the early Earth, we may wonder whether we can really make progress in fathoming the assembly of the first protocells. We can at least begin by looking at modern biological cells, where we see that the most common lipids are a class of molecules known as phospholipids. These molecules consist of two fatty acid molecules that are both attached to a central glycerol (a naturally occurring alcohol) unit, which in turn is decorated with a phosphate, that may in turn be linked to other hydrophilic organic molecules. This general structure is highly suggestive of an evolutionary progression, from primitive membranes composed of fatty acids, to intermediate states composed of a fatty acid joined to glycerol and a phosphate, and then to the modern full phospholipid state. This conceptual framework leads logically to what has been a very fruitful experimental approach, which is to generate small liquid-containing structures known as membrane vesicles composed of these various types of lipids, and then investigate whether their properties are appropriate for the requirements of a primordial protocell that lacks any evolved machinery.

A major difference between modern cells and protocells, considered as compartments with a membrane boundary, is that modern cells have a huge array of evolved protein machines that reside in the cell membrane. This machinery both enables and regulates the transport of everything from water and ions to nutrients and wastes across the membrane. The transporters are complex proteins that are the product of extensive biological evolution, and hence did not exist at the time of the origin of life. In the absence of these protein channels and pumps, membranes composed of phospholipids are formidable barriers to the exchange of molecules between the interior of a protocell and the environment. It is therefore clear that protocell membranes could not have been composed of modern-type phospholipids, even if those had been available as a product of prebiotic chemistry. Note that protocells, by definition, did not contain internal evolved catalysts such as enzymes or ribozymes, so they could not produce building blocks such as nucleotides through internal metabolic processes. This is a critical issue, because it means that the nucleotide building blocks of RNA would have had to find a way to get from the external environment, where they were synthesized, to the interior of a protocell, where they were needed as the elementary units for RNA replication. Based on the impermeability of phospholipid membranes, we can therefore only conclude that primitive membranes must have been quite different. It appears that the early membranes must have allowed for large, polar, and even charged molecules, such as nucleotides, to cross the membrane barrier unaided by any evolved machinery.

Remarkably, it turns out that fatty acids can, and do, spontaneously assemble into classical bilayer membranes in water, with precisely the required permeability properties. The assembly process was first demonstrated over fifty years ago. In fact, such membranes can even form in two distinct ways. If a solution of fatty acids is gently dried down, it forms a thin transparent film composed of many bilayer membrane sheets lying on top of each other like a stack of pancakes. Adding water to this film leads to the entry of water between the membranes, so that the film swells and the individual membrane sheets separate, detach from each other, and eventually close up into vesicles. Alternatively, if acid is added to an alkaline solution of fatty acids, bilayer membranes begin to self-assemble, first as small sheets that grow larger, only to later close up into spherical cell-like vesicles. These vesicles can retain large molecules such as RNA strands indefinitely—a property that ensures genomic RNA molecules will not leak out of their membrane compartment. Fatty acid vesicles that contain encapsulated RNAs are therefore often referred to as model protocells. Notably, fatty acid membranes are much more permeable than phospholipid membranes, to the point that even nucleotides can cross the membranes without help from any evolved protein tools. This surprising property is what makes fatty acid vesicles such great models for the laboratory study of protocells. As we shall soon see, fatty acid membranes have other amazing properties that allow for protocell growth and division.

## Primordial Compartmentalization

The case for a fatty acid composition of protocell membranes is far from open and shut, for two main reasons. First, as we have previously noted, we do not yet have a good understanding of how fatty acids could have been synthesized in sufficient concentrations on the early Earth to allow for the assembly of protocell membranes. For the moment, we can chalk up that deficiency to a lack of sufficient investigation, and hope that a solution will arise in the near future. The second problem is that there appears to be a fundamental incompatibility between fatty acid membranes and RNA copying chemistry. This incompatibility arises because fatty acid membranes are quite sensitive to the presence of common metal ions such as magnesium and calcium. Indeed, fatty acid membranes are destroyed by relatively low concentrations of these ions, both of which are present in many common environments. On the other hand, the replication of RNA requires such ions, typically magnesium, to catalyze the synthesis of RNA. This conundrum is another gap in our understanding of the origin of life. Fortunately, there are a few different ways this problem of compatibility between membrane and RNA chemistry can potentially be addressed. Finding a prebiotically realistic solution to this issue is a hot topic in current research, and we will discuss some of the options being explored later in this chapter.

Still, the lack of a satisfactory pathway for the synthesis of fatty acids, and the apparent incompatibility of fatty-acid-based membranes with RNA copying chemistry, raises the question of whether we should abandon the hypothesis that protocells had a membrane envelope. In other words, wouldn't it be simpler to imagine a protocell without any membrane? This is a very old idea, and the Russian biochemist Alexander Oparin proposed over a century ago that aggregates called *coacervates* were the basis of early cellular life. Coacervates are aggregates of polymers, typically a positively charged polymer (such as a short string of amino acids rich in the amino acid arginine) and a negatively charged polymer, such as RNA. Such oppositely charged polymers attract each other electrostatically, and can form liquid-like droplets that assemble spontaneously in water. These droplets look like cells, and furthermore they can easily take up and release molecules from and to the environment. However, while interesting for their chemical simplicity, such coacervates suffer from two major problems. First, the droplets tend to fuse with each other into larger and larger structures, instead of maintaining their separate cell-like identity. Second, RNA molecules in coacervate droplets tend to exchange rapidly between droplets. This exchange in essence defeats the purpose of having separate cell-like compartments, since no droplet would maintain (at least not for long) a distinct individual identity based on containing a definite set of RNA sequences. Despite these problems, coacervates are the subject of active study and may yet be found to play a role in primitive cells, even if not quite replacing those cells altogether.

Another intriguing alternative hypothesis for primordial spatial localization is that life began with replicating RNA molecules that colonized the surface of mineral particles. In this model, catalytically active RNAs would bind to the mineral surface, and spread across it as they replicated. Rather than the mineral particles growing and dividing like cells, in this model, RNAs would occasionally jump to other mineral particles, spreading and evolving as they colonize additional particles. The model is superficially attractive due to its apparent simplicity, the easy access for surface-bound molecules to nutrients in solution, and the experimental demonstration that activated nucleotides can polymerize on the surface of clay particles. However, it appears that RNA molecules that are stuck to mineral surfaces are distorted in shape by the very forces that cause them to stick to the surface. This shape deformation interferes with replication, and also impinges on the ability of RNAs to fold into functional 3-D shapes.

Although no model of primordial compartmentalization is perfect, in that it is fully supported by both theory and experiments, we will henceforth delve more deeply into the membrane model for two reasons. First, it has been experimentally developed to a much greater extent than any of the alternatives, and second, the membrane model for protocell compartmentalization provides a direct and continuous link to modern biology. Any alternative form of primordial compartmentalization would require a jump to a membrane-based system at some point, which could be quite difficult if functional RNAs had become adapted to a very different environment.

## Genome Assembly

As we have seen, the membrane of a protocell can easily and spontaneously assemble from its component molecules, but the protocell also needs a genome, and at least at first glance, acquiring a genome seems likely to be more complicated, because the assembly of RNA chains from nucleotides requires chemical reactions to connect the nucleotides together. Moreover, this joining of nucleotides into an RNA chain is a dehydration reaction, in the sense that one molecule of water is generated each time two nucleotides link. The condensation of nucleotides into long RNAs, therefore, generates many water molecules, and this process is extremely unfavorable (meaning it requires an input of energy to make it happen) in water as a solvent. Indeed, the reverse reaction, in which water reacts with RNA and hydrolyzes it into its component nucleotides, is much more favorable, which is one of the reasons why

RNA is such a delicate molecule that is so easily degraded. How then can the assembly of RNA, which requires energy, be accomplished in a prebiotically reasonable fashion? Surprisingly, there are several very different ways in which this seemingly difficult task can be realized.

One approach to RNA assembly that has been explored is to simply dry down a solution of nucleotides, either by just letting the water evaporate, or by helping things along with gentle heating. If the nucleotides being used in such experiments have a 5′-phosphate (a phosphate group at the five-position carbon of a five-carbon sugar), then not much happens under mild conditions. However, if the dry-down experiment is conducted with warm carbon dioxide ($CO_2$) blowing over the solution and evaporating the water, some polymerization is indeed observed. This is because some of the $CO_2$ dissolves in the water, forming carbonic acid. Unfortunately, the solution has to become quite acidic for nucleotide condensation to proceed, and as a result of the acidic conditions, the RNA that is formed becomes damaged in a variety of ways. Most commonly, breakage of the bond between a nucleobase and a sugar is observed, generating sites in the RNA chain where the nucleobase is missing. In addition, the RNA sugar-phosphate backbone appears to contain many nonstandard linkages between nucleotides. Consequently, the RNA molecules generated by this simple and plausible but rather harsh process are very heterogeneous, difficult to study, and less than ideal starting points for a protocell genome.

Another possible approach to the initial formation of short RNA oligomers consisting of just a few nucleotides is to begin with nucleotides in which a phosphate group bridges the 2′ and 3′ hydroxyls (-OH) of the ribose sugar. These so-called cyclic nucleotides are the natural products of RNA degradation in water and are also the products of the synthetic pathways developed early on in the laboratory of chemist John Sutherland. Such cyclic nucleotides have long been known to condense (join together) into short RNA chains during dry-down experiments, in a reaction that proceeds at moderate acidity and thus avoids the damaging effects of acid. The condensation reaction in this case is chemically easier, because no water is released during the joining of two nucleotides. The reaction is simply a rearrangement of the phosphate so that it now bridges two nucleotides. This approach also seems simple and plausible, but it does come with some major consequences. In particular, the resulting RNA chains contain many incorrect internal linkages, and always end with the particular structure known as a 2′-3′ cyclic phosphate. Whether this is a bug or a feature depends on your point of view. If RNA can be copied by joining together small fragments with terminal cyclic phosphates on a template strand, this can be a good thing. However, for chemistry that requires a free (i.e., unmodified) 3′-end, this is certainly a big problem. This difficulty could be alleviated if nucleotides are synthesized by a very nonspecific process that adds phosphates randomly to any of the sugar hydroxyls, such that some nucleotides end up with a 5′-phosphate and a free 2′-3′ sugar, while others end up with a 2′-3′ cyclic phosphate, and yet others have no phosphate at all. In this case it may be possible to have the best of both worlds by making short RNA oligomers that terminate in an unmodified 3′-end.

Finally, there is a third approach to the initial assembly of short RNA molecules, which relies on nucleotides with a 5′-phosphate, and special chemistry that "activates" that phosphate to make it much more reactive. Many versions of this so-called activation chemistry have been studied over the decades, beginning with early work by Leslie Orgel. What Orgel and his colleagues found was that chemically attaching a specific type of organic compound (an imidazole group) to the 5′-phosphate of a nucleotide made it possible to observe nucleotide condensation reactions in water. The reason that this works is that imidazole-activated nucleotides release the imidazole when they join together. Imidazole is a good *leaving group* (a group of atoms that detaches from the substrate during the reaction), much better than water in this case, meaning that the nucleotide joining (a.k.a.

condensation) reaction will proceed spontaneously. While these types of nucleotides were developed in order to mediate RNA copying chemistry, it turns out that they can also accelerate random nucleotide condensation in water solutions. Since chemical reactions proceed more rapidly when the reacting compounds are closer together, this untemplated (without a template to direct the joining) reaction in solution is very slow. Not surprisingly, dry-down experiments with activated nucleotides result in much more RNA synthesis. However, the yield of RNA is limited by a competing reaction, namely a reaction with water itself, which reverses the phosphate activation by generating the unactivated nucleotide and free imidazole. There is, however, a simple physical process that simultaneously slows down hydrolysis while increasing the yield of the condensation reaction that generates RNA, namely *freezing*. In a highly counterintuitive process, taking a solution of activated nucleotides that, left to its own devices, would plainly hydrolyze, and placing it in the freezer results in a high yield of RNA! Why does this happen? When water containing dissolved materials freezes, pure water-ice crystals start to grow. As these ice crystals form and get larger, dissolved compounds are excluded from the structure of the ice and accumulate in between the growing crystals. As a result, dissolved compounds become highly concentrated in thin liquid layers between the ice crystals, and due to this concentration effect, molecules that normally might not react at all start to react with each other. This way of generating RNA from nucleotides suffers from the requirement for activation chemistry, but on the other hand, that activation chemistry is exactly what is needed for RNA copying chemistry. What we therefore find is that in an environment that supplies activation chemistry, RNA can be made from nucleotides, then copied, and potentially even be replicated, all stemming from the same chemical process.

By this point we have seen that simple physical processes can lead to the assembly of membrane-bound compartments or vesicles. We have also discovered that a variety of relatively straightforward chemical reactions, coupled with physical processes such as wet-dry or freeze-thaw cycles, can lead to the assembly of short RNA chains. If all of these things happen together, at the same time and in the same place, the result is the assembly of lipid vesicles with encapsulated RNAs. This is, in fact, a routine way of making model protocells for laboratory studies. The question is then whether the same thing could also happen in natural environments on the surface of the early Earth. While it is impossible to know for sure, at least two different geologically plausible environments have been proposed as favorable sites for the assembly of the initial protocells. The first is hot-spring regions, which are widespread and common in volcanically active areas on the surface of today's Earth. Areas of hot springs are likely to have been even more common on the geologically dynamic early Earth. In addition, impact craters host similar areas. In both cases, water circulates through hot fractured rock and re-emerges bearing ions and reactive gases to the surface. Similarly, in both environments, wet-dry cycles can occur readily, as is observed today in the geysers and mud pots of regions around such locations as Yellowstone in the western US. During the winter, freeze-thaw cycles can also arise, and those can even be combined with wet-dry cycles. Consequently, these highly dynamic geological environments may well have provided the necessary combination of conditions for the assembly of the first protocells.

Recently, alkaline carbonate lakes (sometimes called "soda lakes") have been proposed as another, very different type of geological environment that may have also been a favorable home for the first protocells. Such lakes are not common on the modern Earth, but they are still found across our planet. They are typically seen in rather arid regions, in basins with no outlet, where they are fed by groundwater that has filtered through volcanic rocks. This groundwater carries ions leached from the rocks and brings them to the lake, where evaporation leads to concentration and precipitation of salts. These processes lead to an enormous enrichment in dissolved phosphate, a vital component of nucleic acids such as RNA and DNA, and also a key component of phospholipids. For decades, the "phosphate problem" was thought to be a major obstacle for the origin of life, since free phosphate in surface waters is typically only present at extremely low concentrations. This is due to the precipitation of phosphate in the form of highly insoluble calcium phosphate minerals such as apatite. However, this problem is overcome in environments where the concentration of carbonate is very high, as the calcium is precipitated as calcium carbonate (chalk and other minerals such as dolomite), leaving the phosphate in solution. On the early Earth, alkaline carbonate lakes might have provided suitable environments for the synthesis of nucleotides and thus RNA. Alkaline carbonate lakes also provide fluctuating environments, such as wet-dry cycles, resulting from evaporation followed by dilution after rainfall, as well as freeze-thaw cycles during winter. Since the carbonate also precipitates other divalent cations such as magnesium and iron, it turns out that even though the lake water is extremely salty, fatty acid vesicles can assemble in diluted lake water (such as that existing after rainfall) and can survive as clumped-together aggregates, during evaporation-induced concentration.

Hot springs and alkaline carbonate lakes are not mutually exclusive, and a hydrothermally active area that combined both types of geology in close proximity might have provided the best of both environments: reactive derivatives of cyanide from the thermal processing of ferrocyanide salts by lava flows would have been close to the essential sulfurous gases such as hydrogen sulfide and sulfur dioxide from volcanic outgassing, as well as free phosphate in carbonate lakes. Having all of these feedstock molecules together in one place could have set the stage for both RNA synthesis and, possibly, fatty acid synthesis and thus membrane assembly.

There are many details still to be worked out, but the overall picture is gradually coming into focus. The realization that geological environments that are not only reasonable but expected could have supported the necessary chemistry for the origin of life is driving a growing excitement in the field of origin-of-life science, and the feedback between geology and chemistry is accelerating progress in solving the many remaining puzzles.

## On Nonenzymatic RNA Copying

Now that we have at least the outlines of a scenario for the assembly of protocells that look like simplified versions of modern biological cells, that is, membrane envelopes with internal nucleic acids, we are ready to confront the most difficult challenges of all: how such simple protocells, lacking any evolved machinery, could begin to grow, divide, replicate their RNA, and start to evolve. As always, when faced with such a huge and daunting problem, we have to break it down into smaller, more manageable parts that can be tackled more or less separately. Once we achieve a better understanding through this classical reductionist approach, we will be in a position to reverse the process and attempt to build up a holistic view of protocell reproduction. We'll begin by deconstructing the question of how RNA replication could be driven purely by chemistry and physics, without any help from evolved enzymes or ribozymes.

The important feature of RNA is not its chemical composition, but rather the sequence of its nucleotide building blocks. It is the sequence of the A, U, G, and C nucleotides that encodes information, and it is that sequence that must be preserved during replication. RNA replication can be conceptually divided into two phases: first, strands of RNA must be copied, thereby generating a complementary sequence, and second, the complementary strand must be copied to generate a new copy of the original sequence. The basis of the copying chemistry lies in the Watson-Crick-Franklin base-pairing (named after its discoverers, James Watson, Francis Crick, and Rosalind Franklin): U pairs with A, G pairs with C. Therefore, to build up a complementary copy of an RNA strand, all that is needed is to stitch together the correct series of complementary nucleotides. This is precisely what happens in all of modern biology: polymerase enzymes move along a template strand, adding complementary nucleotides one at a time to the growing new strand. Enzymes hugely accelerate this process, as well as making it more accurate, but something very similar can be achieved without enzymes. The basic chemistry underlying this nonenzymatic copying of RNA was worked out by Leslie Orgel and his colleagues and students, beginning in the late 1960s and continuing through the early 2000s. Orgel's key insight was that the substrates used in biology for copying and synthesizing RNA and DNA, namely the nucleoside triphosphates (or NTPs: nucleosides containing a nitrogenous base bound to a five-carbon sugar, with three phosphate groups bound to the sugar), are perfect for enzyme-catalyzed synthesis, but totally inadequate for a nonenzymatic reaction. This is because NTPs are relatively unreactive on their own—an enzyme is needed to speed up the copying chemistry. Therefore, much more reactive nucleotides are required if we are to copy an RNA template without any enzymatic assistance. After trying a wide range of possibilities, Orgel focused on nucleotides that were activated by attaching an imidazole group (possessing three carbon, two nitrogen, and four hydrogen atoms) to the 5′-phosphate. These compounds are indeed reactive enough to start assembling into a complementary strand once aligned on the template by base-pairing. This was a spectacular advance in the early 1970s, and it fueled optimism that the problem of primordial RNA replication would soon be solved. However, that turned out not to be the case, and the limitations of the chemistry soon became apparent: only very C-rich strands could be copied efficiently, and RNAs containing all four nucleotides could not be copied at all. In addition, extremely high concentrations of the imidazole-activated nucleotides were required in order to see any copying, and this was widely viewed as being prebiotically unrealistic. Finally, these reactive substrates hydrolyzed relatively quickly in water, and no prebiotically sensible means of regenerating the activated nucleotides was known at the time. By the turn of the millennium, progress had stalled to the point where Orgel himself and most other researchers in the field felt that the problem of nonenzymatic copying was so difficult that there was little hope for a solution. This discouragement with RNA led to the blossoming of an alternative hypothesis: perhaps RNA was preceded by some progenitor, a simpler genetic material that was easier to replicate, and that was somehow replaced by RNA at a later time in the evolutionary history of life. For most of the 1990s and early 2000s, this hypothesis dominated thinking in the origins community, but in the end, despite a great outpouring of creative and interesting chemical exploration, nothing that worked any better than RNA was found. Then, just as the frustration with finding a path to replication was becoming unbearable, a remarkable turnaround occurred.

It is often said that terrible accidents almost never stem from a single cause, but are instead the consequence of a series of errors. The opposite is also true, and in this case the path to more efficient and general RNA copying chemistry came from the convergence of two distinct advances. The first leap was quite straightforward and had to do with the nature of the nucleotide activating group. In the early 1980s, the Orgel group discovered that a particular substituted imidazole called *2-methylimidazole* was more efficient than plain imidazole as a nucleotide activating group. Even though there was no reason to think that this compound was particularly relevant in a prebiotic sense, it was a good model system. However, no further exploration of the chemistry involved took place for some thirty-five years. In 2016, Li Li, in the Szostak laboratory, undertook a systematic evaluation of the effects of variations in the structure of the activating group, and discovered that changing the methyl group to an amino group greatly improved the rate and extent of RNA copying chemistry. Subsequent work showed that *2-aminoimidazole* (2AI) could be made in two different prebiotically plausible ways. Moreover, 2AI can be made in the same reaction mixture as the closely related compound 2AO, which, as we have noted previously, is a precursor of the nucleotides. 2AI is therefore a reasonable candidate for the actual, prebiotically relevant nucleotide activating compound.

The second advance in RNA copying chemistry was even more surprising than the discovery of 2AI, because the way in which nonenzymatic RNA copying happens is fundamentally different from the way that it happens in biology. Here, the biggest barrier to understanding the correct mechanism was conceptual. The assumption that copying chemistry should work in a similar fashion to the biological paradigm was so deeply ingrained that it was hard to conceive of an alternative. In retrospect though, there were clues to this divergence between the chemical and biological mechanisms in published work from the Orgel lab as far back as the late 1980s. Specifically, in a series of careful and insightful experiments, Orgel's group showed that when only a single activated nucleotide is bound to a template strand next to the RNA primer (a short single-stranded nucleic acid), so that it is poised to react, the reaction occurs very slowly. But when a second activated nucleotide was bound to the template, downstream of the first, the reaction of that first nucleotide with the primer was dramatically faster. In other words, the downstream nucleotide was catalyzing the reaction of the upstream nucleotide with the primer. This mysterious catalytic effect was duly noted by Orgel, but it appears that this effect was never studied any further.

A few decades later, this effect was rediscovered by Noam Prywes in the Szostak laboratory. As the saying goes, spending months working in the laboratory can save you several hours in the library. Once this catalytic effect was firmly established, it became apparent that understanding it could be the key to unlocking fast and efficient RNA copying chemistry. However, figuring out the basis of the catalytic effect took considerable time and effort, and again the slow step was overcoming another mistaken preconception. The initial assumption was that the catalytic effect was due to a physical interaction between the upstream and downstream nucleotides, which served to orient the upstream nucleotide properly for reaction with the primer. Moreover, since both nucleotides had to be activated to see the catalytic effect, the interaction was assumed to be between the activating groups of the two nucleotides. Molecular dynamics simulations showed that the two activating groups could in fact touch each other in several different ways, but not in a way that would obviously stimulate the primer extension reaction. Similarly, crystallography failed to show any stabilizing interaction between the two activating groups. Finally, a critical experiment carried out by Travis Walton, a new student in Szostak's lab, gave the essential clue. The researchers discovered that when two nucleotides were mixed in the presence of the primer template complex, it took almost half an hour for the catalytic effect to become evident. This very long lag time suggested that the two nucleotides did indeed interact, but that this interaction occurred through the formation of a chemical intermediate that took time to accumulate. Soon after this discovery, the anticipated intermediate in the primer extension process was identified by rigorous chemical characterization. It turned out to be an unusual, never-before-seen kind of dinucleotide, that is, a new chemical compound that includes two nucleotides in its structure, bridged by one of the imidazole activating groups. We'll refer to this type of molecule as a bridged substrate.

The role of bridged substrates in RNA copying chemistry was controversial at first. One question that was raised was whether such a molecule could even bind to an RNA template strand by two base-pairs. Doing so would require the bridge to act like a hinge, so that the two nucleotides could lie side by side, yet still be at the right angle to base-pair with the nucleotides of the template. It took structural studies to fully resolve this question, but high-resolution crystal structures clearly showed that bridged substrates could indeed bind to the template RNA strand by two Watson-Crick-Franklin base-pairs.

In a genuine breakthrough, further crystallographic studies revealed the entire stepwise process of nonenzymatic primer extension happening within crystals. In the first step, activated monomers could be seen bound to the template, next to the primer. After some time, the two activated monomers reacted with each other to form the bridged intermediate. And in the final step, the bridged intermediate reacted with the primer, which became extended by one nucleotide, while the downstream half of the bridged intermediate was released. Being able to almost literally "see" the steps of the copying reaction provided three valuable lessons in understanding why the bridged intermediate was so important. First, binding to the template with two base-pairs instead of one explained why primer extension required so much less of the bridged intermediate than monomers—much less was required to saturate the template with the intermediate. Second, the bridge itself was perfectly aligned in 3-D space to favor the reaction with the primer, explaining the faster reaction rate. Third, the new activating group stabilized the structure of the bridged intermediate, thereby allowing it to accumulate to higher concentrations and also probably helping to rigidify its 3-D structure. With this information in hand, further improvements in copying chemistry have come at a faster pace. One such advance came from looking at bridged substrates in which one nucleotide is bridged to a downstream oligonucleotide (instead of another single nucleotide). This provides additional base-pairing to the template, so that template copying can proceed with even lower concentrations of substrates and at even faster rates. On the other hand, competition between different bridged species for template binding seems to slow down overall template copying. Trying to find an optimal compromise between competing factors such as template binding strength, competition, and reactivity is an area of intense ongoing research with the goal of increasing the speed and accuracy of RNA copying chemistry under conditions that are prebiotically realistic.

## Replication

The improvements in nonenzymatic RNA copying chemistry summarized above have brought the next problem into clearer focus. This is the question of how to go beyond simply copying a template to what really matters, which is repeated cycles of *replication*. The problem of replication (for example, copying the copies) is surprisingly difficult, and has a long history of occasional progress, retreats, and roadblocks. Looking at how genomic replication works in biology can give us some sense as to why nonenzymatic replication is so difficult. Biological replication universally involves complex, evolved biochemical machinery, acting in several steps. Replication typically begins at a defined start site that is recognized by specialized proteins. Then highly active polymerase enzymes catalyze the copying of long stretches of DNA (or of RNA in the RNA viruses). Importantly, it is double-stranded DNA (called a duplex) that is copied in cells, generating two daughter DNA duplexes. The way this process works is that the two strands of the parental duplex are pulled apart at a point called the replication fork by special enzymes (*helicases*) that use energy supplied by a molecule known as ATP (which is the energy currency of modern cells), to overcome the base-pairing that holds the strands together. Interestingly, this is the same ATP that is used together with GTP, CTP, and UTP to synthesize RNA in modern cells. As the replication fork progresses, the separated single strands are copied and converted into double-stranded DNA duplexes. Finally, replication ends at special places in the genome where yet more proteins tie up all the loose ends and disassemble the replication apparatus. This tightly orchestrated and highly evolved process clearly developed stepwise from less intricate ancestral systems. How can we gain insight into simpler modes of replication that could give us ideas about how protocell replication might have occurred? One place to look is viral replication, especially the simpler RNA viruses that infect bacteria. These systems have been studied in great detail, but despite their relative simplicity, they also share many features with cellular replication: efficient polymerases that copy long stretches of RNA very rapidly, special machinery for beginning and ending copying at defined locations, and helicase enzymes to separate the strands of a duplex. With none of that evolved framework available at the time of the origin of life, primordial replication must have been quite different and much simpler.

Since we're looking for a less involved approach to replication, what comes to mind is the *polymerase chain reaction*, or *PCR*. This robust technique is widely used nowadays to amplify traces of DNA to useful levels for everything from criminal investigations to the reconstruction of ancient human migrations. Indeed, this Nobel Prize–winning technique has become even more famous due to its use in COVID-19 tests. The essential components of a PCR reaction are simple: an efficient polymerase that is stable at high temperatures, specific DNA primers that define the start sites for the copying reactions, and temperature cycling to separate the strands of the duplex at high temperature while also allowing the enzyme to copy the single strands at a lower temperature. Could thermal cycling have played a role in primordial replication? Instead of using complex evolved enzymes for strand separation, using simple temperature fluctuations does indeed make sense. Temperature fluctuations that are strong enough to drive PCR amplification can be obtained in the same way that heating a pot of water on a stove drives convection—circulation of the water from the hot lower surface of the pot to the cooler upper surface in contact with the air. This, however, still leaves open the question of the primers that define the places where the polymerase starts to copy the DNA. Since a supply of primers with defined sequences is not prebiotically realistic, the replication of a linear genomic RNA sequence can be ruled out. That leaves a circular genome as the logical option, since a circle has neither a beginning nor an end, and therefore it does not matter where copying starts and stops. Turning again to biology for a model, there are certain small RNA parasites of RNA viruses called viroids that use a rolling circle mechanism for their replication. In this mechanism, a polymerase starts copying a circular template, and just keeps going around and around, spinning off a long complementary strand that contains repeats of the circular genomic sequence. An intrinsic ribozyme activity then cleaves the long multimeric (composed of several units) strand into unit-length pieces, and then the same ribozyme ligates the linear pieces into circles, whereupon the process can be repeated. This procedure seems straightforward enough, but it does require a very active polymerase, capable of carrying out a process known as "strand displacement synthesis," and it also necessitates a ribozyme with cleavage and ligation activity. All of this seems like a lot to ask for the very beginnings of replication, long before the evolution of even ribozyme polymerases.

Given the difficulty of identifying a means of replicating an RNA genome without enzymes or other implausible ingredients such as specific primers or ribozymes, we may ask, indeed we should ask, whether there might be a simpler way. Recently, one of us (Szostak) proposed a new model for how primordial RNA replication *might* have worked. Incidentally, this model arose out of discussions held online during the COVID-19 pandemic, when laboratories were closed and experimental work became temporarily impossible. With nothing else to do but think, a means of shedding some of the biologically based preconceptions that had hindered earlier reasoning finally emerged. Following Sherlock Holmes's famous dictum, "when you have eliminated the impossible, whatever remains, however improbable, must be the truth," Szostak and his colleagues tried to piece together the various physical processes and chemical reactions that they thought were reasonable. The idea was to generate a pathway for replication that did not rely on any impossible (or highly unlikely) preconditions. What came up was a model that they (and we here) refer to as the *virtual circular genome* or *VCG* model. This model incorporates the beneficial aspects of a circular genome—that there is no beginning or end and therefore no need to posit the existence of special ways of starting at the beginning and stopping at the end. Instead, copying can begin and end anywhere. However, in the new model the circular genome is virtual, in that there need not be any really circular molecules. Instead, the circle is represented by a collection of overlapping fragments of the circular sequence. The model then makes use of the experimentally well-tested chemistry of template copying by primer extension and by ligation. Since the protocell genome consists of short RNA strands that derive from both strands of the circular genome, overlapping pairs of such oligonucleotides will anneal (base-pair) to form short duplexes with overhanging ends. These overhangs are the sites where primer extension and/or ligation can occur. Thus, in the presence of activated nucleotides and bridged intermediates, small regions of template copying would be expected to occur, at locations that map all over the circle. In the VCG model, copying does not proceed in a directed end-to-end fashion, but is distributed, bit by bit, around the whole circle. Finally, the model invokes environmental variations, such as temperature fluctuations, so that base-paired segments will come apart and then come back together again randomly, with each such cycle shuffling the sets of oligonucleotides that are base-paired with each other. This should enable small amounts of copying to take place at different locations during each cycle. In principle, repeating this process should allow replication of the whole circular set of fragments to occur. We don't know yet whether this can really happen, or if this is just a fanciful scheme that would not work in practice, but intensive theoretical analysis and experimental testing of the model are currently underway.

Despite all the recent progress in working out potentially prebiotic pathways for the synthesis of ribonucleotides and the copying of RNA strands without enzymes, it still makes sense to consider the possibility that life began with some progenitor nucleic acid that later gave rise to RNA. Interest in this possibility received a significant boost when Albert Eschenmoser, one of the greatest chemists to never have received the Nobel Prize (who sadly passed away in 2023), showed that a diverse collection of artificial nucleic acids look like perfectly viable genetic polymers. Eschenmoser designed and then chemically synthesized a range of chemically distinct nucleic acids with unnatural sugars and sugar-phosphate linkages. Surprisingly, many of these polymers exhibited Watson-Crick-Franklin base-pairing, and formed double-stranded duplexes similar to those formed by DNA and RNA. Remarkably, some of these polymers could even base-pair with RNA and DNA, while others formed different groups that could base-pair with each other, but not with members of other groups. These results immediately raised the question of whether any of these artificial nucleic acids could form the genetic basis of alternative forms of life.

How could this possibility be experimentally tested? This is indeed quite difficult to do, since we cannot make and test every possibility. However, we can examine in detail a few special cases that look prebiotically feasible. Two such examples have been looked at in some detail by the Szostak laboratory. These alternative polymers are called *ANA* (*arabinose nucleic acid*) and *TNA* (*threose nucleic acid*), and for both there is a known pathway that makes their presence on the early Earth plausible. ANA differs from RNA only in having the 2′-hydroxyl of the sugar above the ring of the sugar, instead of below, while TNA differs only in that it is missing the 5′-carbon of the sugar. The nucleotide building blocks for both ANA and TNA seem to be likely by-products of the established pathways for the synthesis of the standard RNA building blocks. What would happen if these nonstandard nucleotides were made alongside the usual ribonucleotides in a prebiotic scenario? Experimentally, it appears that both of these alternative nucleic acids are less efficient at template copying chemistry than RNA, so that they lose out in competition with RNA (recall the "RNA always wins" statement). However, RNA strands that also contain these alternative nucleotides mixed in with the RNA can be copied. Remarkably, the copying process preferentially generates RNA. As a result, the outcome of repeated cycles of copying chemistry will be genetic molecules that are almost purely RNA. This is a very satisfying result, since it provides at least a part of an explanation for why biology on Earth began with RNA. However, there is clearly a lot more to be done, as only a limited set of alternatives to RNA have been examined so far.

## Growth and Division

Since two of the most important processes orchestrated by cells are growth and division, we need to return to the questions surrounding the nature of the protocell membrane. We have seen that bilayer membranes can readily self-assemble from fatty acids and other simple lipids. The bilayer structure is similar to that of modern cell membranes, but the properties of the membrane are much more suitable for a primitive protocell than for an advanced cell that runs by using highly evolved protein machinery. A primitive cell has to rely on the properties of its own composition and structure for its survival, so a membrane that spontaneously lets nutrients in and waste products out is essential. The high permeability of fatty acid membranes allows for such transport processes to occur without any evolved channels or pores. Modern cells, however, devote a huge fraction of their genomes to encoding the proteins that mediate growth and division. Could primordial cells really grow and divide just in response to physical inputs from the environment? Surprisingly, experiments show that the answer is a clear yes, and not only that, but there appear to be multiple distinct ways that both growth and division can occur.

One of the most beautiful aspects of the physics of fatty acids is the fact that they are found to adopt such different structures depending on the nature of the environment that they are in. Fatty acids in an acidic solution form oil droplets, while in a basic solution (that has a higher concentration of hydroxide ions than hydrogen ions) they assemble into very small aggregates called micelles. Micelles are only a few nanometers across, and consist of very dynamic assemblies of roughly ten to one hundred molecules. Only at intermediate pH levels, typically very slightly basic, do fatty acids assemble into the bilayer membranes that make up the boundary structure of the protocells we have been discussing. It is the exceptional ability to associate into different phases that provides the basis for a simple means of feeding vesicles so that they can first grow and then divide. When micelles, which are stable only at pH greater than 10 (a basic solution), are added to a suspension of vesicles at a lower pH, the fatty acid molecules in the micelles tend to integrate into the preexisting bilayer membranes, which therefore grow in area. If the vesicles are bounded by a single membrane, the increase in surface area allows the membrane to undergo dramatic shape fluctuations, which eventually lead to the budding of smaller vesicles from the parental vesicle. Thus, merely feeding vesicles with more of their component molecules can lead directly to growth and division, with no need for evolved machinery. A surprising variation on this mode of growth and division occurs if the parental vesicle is a multilayered vesicle. In this case, the outermost membrane layer starts to grow first. Since there is very little volume between membrane layers, the outer membrane initially grows by developing a thin tubular extension, which gradually increases in length and thickness. Over time, the membrane layers exchange material, and the final structure becomes a long multilayered filament. This filament is quite fragile, and mild shear forces, such as those that might be caused by the wind blowing over the surface of a pond, cause the filament to fragment into daughter vesicles. Importantly, the vesicle contents don't leak out during either mode of growth and division, so genetic molecules such as RNA are retained inside the vesicle during the whole cycle of growth and division.

We can now speculate on the type of geochemical environment that could allow protocell growth and division as a result of micelle addition. To start with, we can imagine protocells existing in a pond, stirred by the wind, and held at a slightly alkaline pH by chemicals leached from the surrounding rocks. We then need to envision a separate site, where fatty acids have accumulated, perhaps in a more strongly basic pool, so that they are in the form of micelles. If this reservoir of micellar material overflows, perhaps due to rainfall, a stream of water bearing micelles could flow into the pond containing the protocells, which could then grow and divide. While nothing about such a scenario is grossly improbable, it certainly does seem like a rather contrived set of requirements. So, is there a simpler way by which growth and division might occur? There is indeed an intriguing process by which protocell growth can be driven by competition between protocells. The physical basis of this phenomenon is that vesicles that contain a high concentration of dissolved materials, such as RNA, become osmotically swollen if the surrounding solution contains less dissolved material. In effect, water enters the vesicles, "trying" to dilute the internal RNA to the same concentration as that existing outside the vesicles. Since RNA cannot cross the membrane, the internal pressure builds up, leading to a swollen spherical vesicle. This situation is stable, unless there are other vesicles in the same solution that contain less RNA (or no RNA), and are therefore either less swollen or not swollen at all. In this situation, something stunning happens. Because fatty acids are single-chain lipids, they are not strongly anchored within the membrane, and they can quite rapidly leave the membrane and re-enter either the same or a different membrane. This process allows fatty acid molecules to move between vesicles, and as a result the swollen vesicles tend to grow in surface area as fatty acids enter their membrane. Meanwhile the less swollen or relaxed vesicles tend to shrink as they lose fatty acids. Consequently, vesicles that contain a higher concentration of RNA grow at the expense of neighboring vesicles that contain less RNA. The implication of this mode of competitive growth is that an increase in the rate of RNA replication within a protocell will cause growth of its membrane, at the expense of vesicles in which RNA replication is not occurring or is simply slower. This is a beautiful example of a connection between RNA replication and membrane growth that is based purely on physical phenomena, and not on the evolution of biological machinery. However, there is still a complication. Osmotically driven growth implies that the growing vesicles are swollen and therefore spherical. It is very hard for spherical vesicles to divide since there is not enough surface area to generate daughter vesicles with the same volume. Therefore, division requires a lot of energy to squeeze the spherical vesicles, and results in loss of some of the vesicle contents, unless something else happens to decrease the vesicle volume. It turns out that this could happen easily enough, for example via a sudden influx of salt or other small molecules in the surrounding water. This would cause water to leave the initially swollen vesicle, which would then shrink in volume but retain its original surface area. As above, this would lead to membrane fluctuations, shape changes, and division.

To assemble a fully functional protocell, one more step is needed: we have to combine the very different processes of RNA replication and vesicle replication. Here we are confronted with multiple problems. While each one of these complications has a number of possible solutions, at least in principle, so far these snags remain unresolved. The most immediate and pressing problem is one of systems-level compatibility. The big issue here is the fact that RNA copying chemistry requires relatively high concentrations of divalent cations such as magnesium, but this leads to the prompt and rapid destruction of fatty acid membranes, as mentioned before. So far, two "proof of principle" solutions to this incompatibility have been discovered. The first is to have a so-called chelating molecule bind the magnesium ion, so that one face of the ion is covered up. A very effective example is citrate, which binds magnesium ions quite well, in fact so well that magnesium citrate is a common ingredient in dietary magnesium supplements. The magnesium-citrate complex is still active as a catalyst of RNA template copying chemistry, but interestingly, fatty acid membranes are not affected by the complexed magnesium. Since the membrane is protected from destruction by citrate, it is possible to see template copying proceed on the inside of fatty acid vesicles in the presence of citrate. Although this is certainly encouraging, the prebiotic availability of sufficient citrate is highly unlikely, and other more plausible alternatives are less effective at protecting membranes. It is possible that the mixture of hydroxy and keto acids produced by UV irradiation of lake waters containing carbonate and sulfite might provide partial protection, but other factors must also come into play to resolve this incompatibility. Another partial solution may come from unrelated membrane-stabilizing molecules. For example, recent studies by chemists Sarah Keller, Roy Black, and collaborators at the University of Washington in Seattle have shown that fatty acid membranes are stabilized by components of nucleotides such as ribose and adenine. Membrane stabilization by prebiotically realistic compounds is very promising, and suggests that a combination of these effects with ion-complexing compounds may provide at least a partial resolution to this compatibility problem.

A completely different approach to rendering RNA copying chemistry and membrane integrity compatible is to consider alternative membrane compositions. While fatty acids are key components of modern two-chain phospholipids, intermediates between fatty acids and phospholipids are also prebiotically plausible. The glycerol esters of fatty acids, and the related fatty alcohols, have long been used in mixtures with fatty acids to generate more robust membranes. If we add a phosphate to a fatty acid–glycerol ester, we have a lyso-phospholipid; these molecules are detergents and if present in too high a concentration will dissolve membranes. However, in the presence of phosphate activating chemistry (the same activation chemistry used for RNA copying), that phosphate will cyclize. The resulting cyclo-phospholipids have very interesting properties. Because the cyclic phosphate has only a single negative charge, it interacts only weakly with magnesium ions. As a result, membranes containing a significant fraction of cyclo-phospholipids tend to be resistant to higher levels of magnesium ions. This seems like a potential solution to the compatibility problem, but unfortunately things are not that simple. For protocells to reproduce indefinitely, their membrane composition must be maintained unchanged. However, since cyclo-phospholipids do not form micelles at high pH, it is not possible to feed protocells directly with cyclo-phospholipids. An attractive possibility is that vesicles could be fed with either fatty acids or lyso-phospholipids, which could then be converted into cyclo-phospholipids in the membrane. Other alternatives to fatty acids exist but face similar problems. For example, short-chain phospholipids may be as dynamic as longer-chain fatty acid membranes, given a similar hydrophobic surface area.

The bottom line is that more research into prebiotically reasonable pathways for the synthesis of fatty acids and their alternatives is clearly needed. This is an exciting research frontier, because the prize is an explanation for how populations of protocells could be maintained, while circumventing the incompatibility between RNA chemistry and membrane properties.

We have in this chapter deliberately given a quite detailed account of the experimental landscape. Our goal has been to demonstrate that by considering the physical processes of membrane growth and division, and the chemical processes of RNA replication, we have come to a reasonable, albeit incomplete, model of the nature of the first cells on the early Earth. This scenario is by no means the last word, and much remains to be done to fill in the missing details. Nevertheless, the current model is based on a considerable body of experimental work, and it is supported by sound theoretical arguments. If we are willing to accept, for the time being, the broad outlines of the protocell model, we can begin to think about the evolutionary path that led to the development of the key features of modern cells that were present in the common ancestral cell from which life originated (the Last Universal Common Ancestor, or LUCA). These features include cell membranes made from more complex lipids and the wide array of embedded proteins that control molecular transport in and out of all cells. A second key feature is a complex network of metabolic reactions that generates many if not all of the cell's nutrients. Finally, there is RNA-encoded protein synthesis, which allows for the production of protein enzymes, membrane proteins, and the cytoskeleton (the structure that helps cells to maintain their shape and functions). Here we will briefly sketch out a possible scenario for the evolution of these hallmark features of biology.

One particularly attractive model for how protocells might compete for a limited resource involves the ribozyme-mediated synthesis of two-chain lipids similar to modern membrane phospholipids. Experiments in the Szostak laboratory have shown that the presence of a small amount of such lipids in the fatty-acid-based membranes of model protocells allows such vesicles to grow by stealing fatty acids from neighboring vesicles that do not contain phospholipids. This effect arises because fatty acid molecules are held more tightly in membranes that contain even a small fraction of two-chain phospholipids. As a result, fatty acid molecules escape more slowly from the membrane into the surrounding solution, but since they enter the membrane at the same rate, the membrane grows while the surrounding pure fatty acid membranes shrink. An important implication of this physical effect is that a protocell that happened to evolve a ribozyme that could catalyze the synthesis of phospholipids could grow at the expense of its neighbors. This competitive advantage would help these phospholipid-making cells to take over and dominate the local protocell population. However, once the descendants of that original cell have taken over the population, their selective advantage would wane, since all cells would be making phospholipids. At this point, an interesting evolutionary arms race would begin. The reason is that any cell making more phospholipids than its neighbors would still have a competitive advantage and could grow by absorbing fatty acids from its neighbors. Such a competition would lead to a gradual increase in the efficiency of the phospholipid-synthesizing machinery, and thereby also to a rise in the abundance of phospholipids in cell membranes. However, an increasing fraction of phospholipids would soon have powerful effects on cell physiology, both positive and negative. Fatty acid membranes with a significant fraction of phospholipids start to become less permeable to polar and charged solutes. Thus, a protocell that survived by taking up nutrients such as nucleotides from its environment would no longer be able to do so, or rather, that nutrient uptake would become progressively slower, as the membrane phospholipid content increases. On the other hand, reduced membrane permeability opens up new possibilities for the evolving cell, because internal metabolic reactions could finally become advantageous, since molecules synthesized inside the cell would no longer leak out quickly and feed the neighboring cells. Instead, internal metabolic pathways could benefit the cell housing the ribozymes that catalyze the metabolic reactions. Alternatively, cells could begin to evolve specialized systems for the import of needed nutrients. Reduced membrane permeability would therefore lead to two new selective pressures, one for the evolution of intracellular metabolic reactions, and the other for the evolution of membrane transport machinery.

The evolution of metabolism must have proceeded step by step, with each stage providing a selective advantage. A reasonable way this could be achieved is through the evolution of new ribozymes, each catalyzing the synthesis of some compound that the cell is running out of or is having difficulty in accessing. Since new ribozymes would be encoded by the cell's RNA genome, this evolutionary progression likely proceeded in stages. An early requirement for an expanded genome that could encode additional ribozymes might have been increasing the rate and accuracy of genomic replication. As environmental sources of nucleotides were depleted, increasing the internal synthesis of activated nucleotides from progressively simpler and more readily available feedstock molecules would have been beneficial. How far this process could go with only RNAs as catalysts is highly uncertain. What is clear is that at some point, life evolved the ability to synthesize useful peptides and then gradually larger and more active enzymes. However, this advance required the emergence of coded peptide synthesis, and the evolution of the genetic code remains one of the great mysteries of early evolution. That process is also thought to have occurred over many steps. Some of the most abundant and easy-to-make amino acids might have been the first to be included in the genetic code, while amino acids with a more complex biosynthetic pathway might have been integrated into the genetic code at later times. This hypothesis is consistent with the fact that the easiest-to-make amino acids tend to have the strongest codon-anticodon pairing (between the three-nucleotide units of genetic code in transfer RNA and messenger RNA), and also tend to occupy the so-called family boxes of codons, that is, the different groups of four codons that all code for the same amino acid. Beyond that, the relationship between amino acids and codons seems to be arbitrary, so that the genetic code as a whole can be viewed partly as the result of a deterministic pathway, and partly as the result of a frozen historical accident.

This brings us finally to the topic of the next chapter, the connection between the origin of life on Earth and the formation and evolution of the Earth itself.

## CHAPTER 6

# Putting It All Together

## From Astrophysics and Geology to Chemistry and Biology

*Forces of nature act in a mysterious manner. We can but solve the mystery by deducing the unknown result from the known results of similar events.*

—Mahatma Gandhi, *Soul Force*

We know from isotopic dating that Earth formed some 4.54 billion years ago, so to fathom much about what the early Earth was like presents a serious challenge. Fortunately, in addition to Earth's own geological record and information obtained from the study of other objects in the solar system (Mars and Venus in particular), we can also gain insights into how Earth was born through astronomical observations of many stars around which planets are still in the process of being formed. Via the stunning multi-wavelength images and detailed spectra provided by such observational facilities as the Hubble Space Telescope (HST), the James Webb Space Telescope (JWST), and the ground-based Very Large Telescope (VLT), in addition to incredibly rich data from the Atacama Large Millimeter/submillimeter Array (ALMA) radio observatory, we can literally witness how vast molecular clouds of gas and dust condense under the force of gravity to form dense protoplanetary disks around nascent stars. Those disks become the nurseries inside which planets are born. In a growing number of cases, astronomers have even used spectroscopic observations to map the chemical composition of these circumstellar disks. In fact, researchers have gone even further. When planetary and atmospheric scientists, astrophysicists, and geologists combine the available wealth of data with sophisticated computer simulations, they are able to produce a dynamic, if only partial, picture of how planets form. The infrared vision of JWST offers striking new capabilities. By being able to peer through some of the otherwise opaque (to visible light) cosmic dust, it is already providing us with an unprecedented view of the birth of stars and planets. In particular, the combination of observations from ALMA and JWST covers both planets that form farther out from their host stars and those that are born closer in.

There are still many questions concerning the early history of planet Earth, however, that cannot be answered simply by looking at other solar systems. A central question, for example, has to do with the precise history of asteroid impacts on Earth. Asteroids are rocky remnants left over from the early formation of our solar system. There are more than a million known asteroids, of which about six hundred thousand have well-determined orbits. Most of this space rubble is orbiting the Sun between Mars and Jupiter, where Jupiter's gravity prevented larger bodies from forming, creating what is known as the main asteroid belt. Observed asteroids range in size from about 584 miles (940 kilometers) in diameter (for the asteroid Ceres, classified as a dwarf planet), to objects that are less than about 33 feet (10 meters) across.

From the fact that Earth and the Moon have identical isotopic compositions, we know that once Earth was almost at its full size (about

4.42 to 4.52 billion years ago), it was blasted by a cataclysmic impact of a Mars-size body (dubbed *Theia*), which tore away a part of Earth and led to the formation of our Moon from the debris. At the same time, this collision melted Earth's outer layers, creating a magma ocean that probably took a few tens of millions of years to cool down and solidify. That much is virtually certain. Collisions may have also altered Venus's spin around its axis (it takes Venus longer to spin on its axis than to orbit the Sun). Uncertainties start creeping in when it comes to the details of subsequent impacts on Earth's surface. On the basis of dating studies of craters on the lunar surface, researchers concluded that the early Earth must have been bombarded by many more large impacts. However, the precise size distribution of the impactors and the rate and timing of the impacts remain subjects of heated debates, as do the effects of those impacts on the young Earth. While a few scientists proposed that roughly between 4.1 and 3.8 billion years ago Earth was subjected to what has been called the "Late Heavy Bombardment" (LHB)—a disproportionately large number of asteroids and comets (the latter being snowballs of ices, dust, and rock) colliding with the early Earth—others maintain that the LHB never happened. The putative evidence for the LHB came from rock samples brought back from the Moon by Apollo astronauts. Isotopic dating of those rocks suggested that they were molten (through impacts) during a rather narrow time interval. Those who are skeptical about the reality of the LHB have argued that the apparent sharp peak in the ages of lunar impacts is a statistical fluke, produced by the astronauts having in fact sampled rocks scattered from a single large impact. These scientists suggest that instead of a late heavy bombardment, there was a brief early heavy bombardment that ended around 4.4 billion years ago, after which Earth continued to experience a long period of declining bombardment for about two billion years that continues at ever lower rates until today. The reason that these details are important is that they have important consequences for determining when Earth became habitable, and when life could have emerged.

Some of the giant impacts on Earth might have been sufficiently catastrophic to have melted a part of Earth's crust, destroying most if not all of budding life. On the other hand, somewhat less powerful impacts might have had the opposite effect. The collisions could have temporarily created a reducing atmosphere (containing reducing gases such as hydrogen and hydrogen cyanide), which, as we have seen in Chapter 3, would have been more conducive to the origin of life, since cyanide would have been an excellent starting material for the chemical synthesis of the building blocks of life. Asteroids could have also brought metals, such as iron, and other elements, such as phosphorus and sulfur, that are necessary for life. Even most of Earth's water (perhaps as much as eight times the total amount in all the oceans on Earth's surface today, according to one simulation) may have been delivered from early impacts. Much smaller amounts of prebiotically relevant compounds, such as amino acids (chains of which could form proteins) and nitrogen-containing bases of the types that compose RNA, would also have been delivered from certain types of meteorites. Even smaller impacts would have generated craters with associated geothermal activity leading to regions of hot springs and mineral-enriched ponds and lakes that, as we have described earlier, could have been ideal local environments for the emergence of life. When it comes to life, therefore, it is definitely the case that asteroids giveth and asteroids taketh away. Just as an amusing aside, in 1992 Michelle Knapp of Peekskill, New York, had bought a car for $300. After the car was totally destroyed by a small meteorite impact, she sold it to a collector for $25,000!

The early history of Earth provided astrophysicists with yet another puzzle, known as the "faint young Sun paradox." This "paradox" arose from the discovery (in the Jack Hills in Western Australia) of tiny crystal minerals known as zircons, dating back in one case to about 4.4 billion years ago. An analysis of the ratio of different oxygen isotopes in the zircons showed that even at those extremely early times, liquid water existed on or near Earth's surface. The paradox refers to the apparent contradiction between the expectation from stellar evolution models, which tell us that the Sun's luminosity at that period was about 30 percent lower than its current value, and the evidence for the presence of liquid water at the very same time. For such a faint Sun, the radiative heating it provided was low enough to suggest that all surface water on Earth should have been frozen solid, and life could not have emerged and developed.

While the precise solution to this "faint young Sun" puzzle is still not known with certainty, the general thought is that a combination of several effects might have been at play. In particular, the presence of higher concentrations of greenhouse gases such as carbon dioxide (and possibly some methane) in Earth's early atmosphere might have prevented a complete freeze. Carbon dioxide would certainly have been released from volcanos. Suggestive evidence for this scenario comes from an analysis of the composition of meteorites dating to 2.7 billion years ago. The data showed that as these meteorites descended through Earth's atmosphere, they interacted with a carbon-dioxide-rich mixture of gases (with carbon dioxide making perhaps as much as 70 percent). Other contributing factors that could have helped in keeping Earth from freezing were as follows: (i) Earth might have reflected less energy back into space, (ii) asteroid impacts, the rate of which would have been high 4.5 billion years ago (irrespective of whether or not the Late Heavy Bombardment truly happened), could have generated enough heat to melt the frozen water at least episodically, and (iii) heat could have been generated by pushing and pulling in Earth's interior, caused by the tidal forces exerted by the Moon, which was much closer to Earth shortly after its formation.

The problem posed by the "faint young Sun paradox" is exacerbated on Mars, since data from the *Perseverance* rover and other robotic missions indicate that lakes of liquid water and flowing rivers existed on the Martian surface 3.7 billion years ago or even earlier. This puzzle will perhaps be resolved only when samples gathered by *Perseverance* are returned to Earth for analysis in the 2030s. The "faint young Sun paradox," however, presents what is perhaps an even more intriguing possibility, with respect to both life on Earth and extraterrestrial life. Atmospheric scientist Martin Turbet of the Dynamic Meteorology Laboratory in Paris suggested in 2021 that had our Sun been at 92 to 95 percent of its present luminosity some 4.5 billion years ago, water vapor in Earth's atmosphere would not have condensed to form liquid water. Instead of a "snowball Earth" we would have had a "steam Earth." In other words, the faint young Sun may have been a blessing rather than a curse—a necessary condition for life to emerge on Earth! In contrast, on Venus, for example, Turbet's models predict that the planet was never sufficiently cool to sustain liquid water. The good news is that this particular prediction may soon be tested. NASA plans to send two spacecraft to Venus this decade, and the European Space Agency will also send one. One of the NASA missions, called *DAVINCI* (short for Deep Atmosphere Venus Investigation of Noble gases, Chemistry, and Imaging), will include a probe that will descend into Venus's atmosphere and determine its composition at different heights. The chemical signatures obtained in this way will provide meaningful clues for establishing whether past oceans existed on Venus. In addition, surface mapping results that will be collected by the other two planned Venus orbiters—NASA's *VERITAS* and the European Space Agency's *EnVision*—will allow for the calibration and testing of Venusian climate models. We shall further examine Venus as a potential candidate for harboring life in Chapter 7.

The bottom line of this brief discussion is clear. From a fusion of findings from the geological and atmospheric history of Earth, exploration of other solar system planets, observations of exoplanetary systems as they are being formed, studies of meteorites impacting Earth, numerical simulations, and more recently even data from a few sample-return missions (from visits to asteroids and comets), we can attempt to construct a piece-by-piece picture of how Earth formed, and the initial conditions from which life emerged.

Then, however, we hit another snag—the obstacle of Earth's so-called missing years. Here is the problem. On Earth, the outermost rocky shell (the *lithosphere*) is cracked and broken up into large pieces, known as *tectonic plates*. There are seven major plates and ten minor ones. Since the outer part of Earth's core is molten (making the hot mantle underneath the surface crust partly fluid), these tectonic plates slowly move about, sometimes colliding, or even diving underneath one another into the subsurface, in a process known as *subduction*. Because of plate tectonics, large chunks of Earth's surface have been continually recycled by subduction—one chunk slides beneath another and sinks down until it has melted. Over billions of years, this activity has erased all signs of Earth's very early history and hence of precisely the environments we are most interested in—those that gave birth to life. Re-creating, in our mind's eye, those prebiotic conditions therefore requires a careful integration of several lines of evidence. Those include an extrapolation from processes we can still observe on the modern Earth, the results of rigorous computer modeling, and in particular, detailed studies of the Martian surface. What makes studying Mars so crucial is the fact that on Mars much of the record that is missing on Earth has been preserved (since, as we shall see in the next chapter, Mars lacks plate tectonics). The successful landing of the *Perseverance* rover on Mars is very promising in this respect. A meticulous weaving together of all the information gathered by *Perseverance* and previous Mars landers and orbiters is helping us to shed light on the distant past, while at the same time highlighting the remaining unsolved puzzles that are driving current cutting-edge research.

Moving on from geology to the chemistry frontier, an important lesson that we have often discovered and rediscovered is that progress can be hindered by our own preconceptions, and breakthroughs sometimes have to wait for a new way of thinking to take hold. Humorously, this fact is sometimes called "Planck's Principle," because physicist Max Planck, the originator of quantum theory, wrote: "A new scientific truth does not triumph by convincing its opponents and making them see the light, but rather because its opponents eventually die and a new generation grows up that is familiar with it." One example, which we have already encountered, that demonstrates the hindering effects of misconceptions is related to the intertwined relationship between prebiotic chemistry and early, only half-baked ideas about the environments that were required for that chemistry of life to materialize. Specifically, it concerns one of the most famous experiments in the history of origin-of-life studies—the *Miller-Urey experiment.*

While there is no doubt that the Miller-Urey experiment, which produced molecules closely related to amino acids from a relatively simple chemical setup, represented a major breakthrough, if we examine more closely the actual outcome of that demonstration, we realize that very small amounts of thousands, even tens of thousands, of different chemicals were created. This is definitely not what we would like to have in order for life to get started. Rather, what we need are high concentrations of just a few key chemicals. Clearly something very important was missing or simply wrong. Significant advances in the last twenty years (which we have described in detail in Chapters 3 and 4) have uncovered a blueprint for such high-yielding specific chemical pathways. But this improved chemistry has its own very particular requirements. For instance, we explained how it is driven by energy provided by UV radiation. On the one hand this sounds promising, since we know that young stars emit copious fluxes of UV light. On the other, it raises a problem: such energetic radiation might have rapidly destroyed the very molecules we need in order to build life. Surprisingly, it turned out that a combination of observations, experiments, and modeling has given us an astonishing answer: the most potentially destructive high-energy UV radiation would have been filtered out by Earth's atmosphere. In contrast, the moderate-energy UV light, the very UV that could stimulate the right synthetic reactions, could get through the atmosphere, and with intensities that would have been sufficient to expedite the chemistry needed for life.

This important realization was an excellent first step, but it also meant that the required UV light could only reach Earth's surface if it wasn't blocked by a thick haze of hydrocarbon particles of the type that, for instance, shrouds Saturn's moon Titan. Such a haze would have definitely covered the early Earth had its atmosphere truly consisted of hydrogen, methane, ammonia, and water, as was generally assumed at the time of the Miller-Urey experiment. More recent research indicates, however, that Earth's early atmosphere had an entirely different mixture of gases, with the most likely composition being primarily a mixture of carbon dioxide and nitrogen, with only trace amounts of hydrogen and other minor gases. This, however, leads to another problem: it is very difficult to synthesize more than a trace amount of cyanide (crucial for the emergence of life) in an atmosphere dominated by such stable gases. A recent model suggests that there is a way in which we might have our cake and eat it too: as we discussed above, moderately large asteroid impacts could have generated a transiently reducing atmosphere dominated by hydrogen, nitrogen, and methane. While this more reactive atmosphere lasted, large amounts of cyanide could have been produced, then captured and stored on the surface as ferrocyanide salts. Subsequently, once the reducing atmosphere had dissipated and the atmospheric composition was once again dominated by carbon dioxide from volcanic outgassing, any haze that might have formed would also have disappeared, and the necessary midrange UV could once again reach Earth's surface. This remarkable consistency, between the advantages of photochemistry, the spectrum of the radiation emitted by the young Sun, and the likely nature of early Earth's atmosphere, suggests that at the very least we should keep this promising framework in mind, while continuing to search for additional supporting evidence. We should also be striving for such self-consistent scenarios while exploring the possibility of life on other solar system bodies and on exoplanets.

This seemingly self-evident idea—that the planetary environment and the prebiotic chemistry that together led to the emergence of life have to be mutually compatible and consistent—serves as a powerful constraint on any model. In addition, the major conceptual step forward has been the recognition that an analysis of the chemistry can tell us something important about the environment, and a critical examination of potential environments can teach us about the required chemistry. It is precisely the exploration of the surprising connections inferred in just this way that has produced a broadly comprehensive view of the early Earth, and has led to what we currently conjecture might have been the birthplaces of life.

## The Emergence of Life

If we knew for sure when primitive life existed on the early Earth, we could begin to narrow down the window of time when life first emerged from the chemistry of our young planet. Consequently, the search for evidence of Earth's earliest life is an ongoing effort, though extremely challenging. This difficulty is partly because very little of Earth's ancient crust remains, with almost all of Earth's oldest rocks having been destroyed by subduction. Moreover, the few remaining areas of very old terrestrial surface have been heavily modified by eons of heat and pressure, shattering what little evidence was held in those rocks. Even worse, a surprising variety of non-biological processes can lead to microscopic mineral structures that look remarkably lifelike. Just because something looks like a fossilized form of microbial life doesn't mean that it really is. Great care and caution are required when interpreting the formations found in ancient rocks. What then is the oldest clear-cut evidence for life on Earth?

Arguably the best and most solid (so to speak) evidence of early microbial life is that preserved in hard, layered structures known as *stromatolites*, which are created by microorganisms. Stromatolites were (and are) formed when sticky mats of microbes trap and bind sediments into layers. Subsequently, as minerals precipitate inside those layers, they produce durable arrangements with a characteristic layered structure commonly in the shape of rounded or conical domes. Stromatolites can be observed today, growing in shallow-water marine environments. Very well-preserved fossilized stromatolites as old as 2.5 billion years are found in several locations on Earth, but the very oldest fossilized stromatolites are found in Western Australia. These date to about 3.4 to 3.5 billion years ago, and have been intensively studied for decades by Australian astrobiologist Martin Van Kranendonk and his colleagues and students. Because of these careful and thorough studies, we can be confident that life was widespread in shallow marine environments less than a billion years after the formation of Earth. A billion years, however, is a long time, almost a quarter of our planet's history. Can we do better in pinning down when we are sure that life existed?

Unfortunately, all efforts to obtain definitive evidence for microbial life at earlier times remain controversial at best. Cell-like microfossils with walls composed of organic materials containing carbon and nitrogen have been found in ancient rocks in Australia, Greenland, and South Africa. But whether the oldest of these fossils (3.4 to 3.7 billion years old) are truly fossilized microbes or simply non-biological structures remains unclear. Similarly, formations that were initially claimed to be microfossils possibly older than 4.3 billion years, in an ancient hydrothermal vent in the Nuvvuagittuq Greenstone Belt of Quebec, Canada, based on filaments and tubes made of a type of rust encased in layers of quartz, are now generally believed to be abiotic (unrelated to life) mineral structures. Another claim for evidence of very early life was based on carbon-isotope ratios from graphite inside a zircon crystal as old as 4.1 billion years. However, this graphitic carbon is now thought to derive from a non-biological source. Although none of these claims for very early life are generally accepted, it should be noted that every such effort has pushed forward the technology being used to characterize these microscopic ancient structures. We can certainly hope that further advances in technology will one day provide us with definitive evidence concerning the age of the earliest life.

A completely different method by which researchers have attempted to deduce the time at which microbial life was present on the early Earth involves the concept of a "molecular clock." By comparing the genomes of modern organisms and assuming (a big assumption!) that neutral mutations happen at a constant rate, it is possible to surmise the existence and age of the Last Universal Common Ancestor (LUCA). LUCA is defined as the most recent common ancestor of all current life. Because LUCA shares the common features of biochemistry found in today's bacteria, archaea, and eukaryotes (the three domains of cellular life-forms), LUCA is inferred to have had a complexity similar to that of modern microbes, for example, a DNA genome, ribosomes for protein synthesis, many protein enzymes, a complex metabolism, and evolved cell membranes. There is therefore a huge evolutionary distance between LUCA and the first protocells. The time required to go from the first simplest cells to LUCA is completely unknown, and this remains a major puzzle. While efforts to use molecular clocks have typically given very early dates for LUCA (more than 4 billion years ago), molecular clocks can suffer from considerable uncertainties, due to the fact that the rate of evolution can vary greatly. In particular it might be expected that the accuracy of DNA replication might have been lower at very early times, which would result in higher error rates and a molecular clock that ran much faster at earlier times. This would give a false impression of the antiquity of ancient life. Therefore, unless and until new and better evidence is obtained, we are left with a great uncertainty concerning the timing of the origin of life.

In addition to the question of *when*, origin-of-life researchers have long debated the question of *where* on Earth life first emerged. The first modern suggestion of a suitable location and environment for the origin of life was Charles Darwin's "warm little pond, with all sorts of ammonia and phosphoric salts, light, heat, electricity, etc. present." Despite the fact that Darwin knew nothing of the role of nucleic acids in heredity, and nothing of the chemistry leading to nucleotides, amino acids, and lipids, this suggestion was prescient in many ways. Small ponds allow the ingredients of life to be concentrated by evaporation, and allow UV light to drive the essential chemistry forward. This model has become much more sophisticated over the subsequent century and a half, as the warm little pond is now seen in a broader context of volcanic or impact-driven hydrothermal activity, as we shall discuss in more detail below.

Before we return to the geophysical and geochemical context of that natal pond, however, we should first discuss the other model for the site of life's origin, namely the deep-sea hydrothermal vents model that has been so widely discussed in the popular press. The motivation for considering deep-sea hydrothermal vents as a suitable location for the origin of life stems from the fact that these are locations that have been dramatically and beautifully colonized by modern life. The high-temperature vents are located at deep-sea sites where new crust is being formed as magma rises, cools, and fills the gap between tectonic plates that are being pulled apart. These are sites where abundant energy is available in the form of redox gradients (a series of reduction-oxidation reactions), which result from the circulation of seawater through the fractured rock, and which bring reduced metal ions to the surface of the rock in plumes of hot water. Once released into the large volume of water, these metal ions react with oxygen in the seawater and precipitate in clouds of metal oxides and hydroxides, giving rise to the term "black smokers." The very different low-temperature, off-axis vents release plumes of alkaline water into the more acidic seawater, again providing a source of energy that is efficiently harvested by the abundant life found at these vents.

However, just because a particular environment has been colonized by modern life does not mean that this environment is suitable for the *origin* of life. Let us therefore briefly consider the model proposed for a deep-vent origin of life. Not long after the discovery of the alkaline vents, British geologist Michael Russell and microbiologist William Martin of the Heinrich-Heine-Universität in Düsseldorf, Germany, proposed that microscopic pores in the chimneys of the alkaline vents could have played the role of early protocells, allowing metabolism to start, and eventually developing into actual cells. In this "metabolism first" model, simple chemical reactions catalyzed by metal ions lead to the emergence of more complex metabolic pathways, which eventually "invent" routes for the synthesis of nucleotides, amino acids, and lipids, finally transitioning to cells with nucleic acids for heredity and proteins for catalysis. Unfortunately, this model is fundamentally flawed from a chemical perspective. First and foremost, there is no chemically realistic way for metabolic pathways and cycles to develop in the absence of powerful catalysts such as RNA or protein enzymes, which of course are encoded by nucleic acid genes that in turn emerged from the process of Darwinian evolution. Despite decades of discussion, no evidence whatsoever has emerged of any prebiotically plausible chemical reactions occurring in deep-sea hydrothermal vent environments that might conceivably lead to nucleotides and RNA, or amino acids and peptides, or lipids and membranes. In the absence of the basic chemistry required for the synthesis of the building blocks of life, it is a fundamental misconception to view deep-sea hydrothermal vents as the site for the origin of life. At this point it is almost superfluous to list all of the other flaws in this model, but for completeness we will still do so. First, the chemical reactions via which chemist John Sutherland and his collaborators managed to create RNA nucleotides and amino acids (as described in Chapters 3 and 4) require geological settings that allow for the *concentration* of starting materials. This is easy on Earth's surface, where simple physical processes such as evaporation, crystallization, and freezing can lead to the concentration of key starting materials and intermediates that cannot be achieved in a large ocean, where the chemicals would be diluted and lost. Second, many of the processes that are crucial for the origin of life have been found to be powered by UV radiation, which implies that they couldn't have taken place at the bottom of the ocean. Third, complex and varied surface environments would allow for different chemical processes to occur in different locations and at different times, combining at later times to enable subsequent steps on the overall path to life to happen. In other words, the origin of life had to occur on land, in small lakes or ponds, where the Sun could both provide the required UV input and produce (aided by geothermal heat) wet-dry cycles, freeze-thaw cycles, and the accumulation of reservoirs of intermediates. All in all, the deep-sea hydrothermal vent hypothesis for the origin of life is a classic example of how a deep-seated misconception, derived from an interesting initial finding but held on to in spite of accumulating contradicting evidence, can hold back the progress of a field by diverting attention and resources from more realistic scenarios.

Resurfacing, so to speak, let us consider how varied environments, as diverse four billion years ago as today, could have facilitated the emergence of life. An important point to bear in mind is that getting from a lifeless planetary surface to even the simplest early protocells is not a one-step process. Many components have to come together to form and maintain life, and these different components would not all be made or found in the same place at the same time. How the different ingredients of life were produced, accumulated, and then assembled remains a fascinating puzzle, but we can now begin to see bits and pieces of the overall picture. Alkaline carbonate lakes are a far cry from Darwin's warm little pond, but they do have properties that make them ideal for at least two of the early steps in getting to life. Careful modeling by University of Washington astrobiologists Jonathan Toner and David Catling first showed that such lakes could accumulate ferrocyanide complexes over long periods of time, capturing cyanide that was rained out of the atmosphere in dilute solution, and allowing reservoirs of cyanide to accumulate. The same lakes could also accumulate phosphate in a soluble form, so that it would be available to act as a pH buffer, an acid-base catalyst, and to be incorporated into the nucleotide building blocks of RNA. Needless to say, none of these processes could take place in the oceans. Accumulating ferrocyanide in sediments is an efficient way to build up a reservoir of the key starting material cyanide, but the ferrocyanide salts need to be thermally processed in order to allow for a richer chemistry to emerge. In a volcanically active region, this could easily be achieved by lava flowing over a dried bed of ferrocyanide-rich sediments. The heat and pressure from a meteorite impact could have similar effects, releasing the cyanide from its iron grip, and transforming some of it into energy-rich compounds such as cyanamide and other cyanide derivatives. Subsequently, rainwater percolating through cracked rocks could dissolve these soluble high-energy compounds, resulting in groundwater that was saturated with reactive feedstock molecules. This groundwater could then transport its load of starting materials into ponds or lakes, more closely resembling Darwin's warm little pond, which, just as he said, would be filled with phosphate and the carbon-nitrogen compounds needed to bootstrap life. When in a surface environment exposed to UV light, and with access to volcanic gases such as hydrogen sulfide and sulfur dioxide, the "cyanosulfidic photoredox" chemistry that we have described in detail in Chapters 3 and 4 could commence, leading to the synthesis of nucleotides and amino acids.

Once nucleotides, amino acids, and related compounds are present at high concentrations in localized surface environments, there are many ways in which physical processes such as wet-dry and freeze-thaw cycles can contribute to the assembly of larger, more complex molecules. For example, activated nucleotides can link together into short RNA chains by partial drying, which generates a slurry or paste in which the nucleotides are so concentrated that they start to react with each other to form short oligomers. Similarly, activated nucleotides dissolved in water will not polymerize, but if that solution freezes, as might happen during a winter cold snap, polymerization starts to happen because the nucleotides become concentrated in the thin liquid zones between the growing water-ice crystals. Wet-dry cycles can also lead to the formation of peptides. In one interesting process, alpha-hydroxy acids, when dried down, spontaneously react with each other to form polymers known as polyesters. Amino acids can then attack these ester linkages, becoming incorporated into a mixed polymer of amino acids and hydroxy acids. Continued wet-dry cycles in the presence of amino acids will then eventually generate short peptides. Another fascinating possibility is that cyanide dissolved in water, in the presence of sulfide, would hydrolyze to make formamide, which is much less volatile than water. Therefore, the evaporation of the water could leave behind a liquid composed mostly or perhaps entirely of formamide, which happens to be an excellent solvent for a number of important organic reactions. Finally, we have previously discussed the fact that both RAO, the intermediate on the path to nucleotides, and CV-DCI, the reservoir form of cyanoacetylene, crystallize beautifully from water, potentially allowing the accumulation of purified reservoirs of these important compounds in surface deposits. *The important point is the fact that all of these linked physical and chemical processes can only occur in surface environments, which provides strong circumstantial evidence that the origin of life itself was a surface phenomenon, and could not have happened in the oceans in deep-sea hydrothermal vent environments.* The general lesson here is that models that neglect the multistep chemistry required to produce and accumulate the key building blocks of life are not realistic models of the origin of life and do not help to advance our understanding. In contrast, models that embrace and explore the difficulties of multistep chemical pathways lead to insights into relevant geochemical environments, and the consideration of relevant surface environments in turn leads to insights into realistic chemical processes that could have contributed to the origin of life.

If we now consider higher-level processes such as nonenzymatic RNA replication, and the growth and division of protocells, we find once again that a dynamic, fluctuating environment is almost certainly a necessity for the propagation of early life. We can see why this is the case if we compare the processes of enzyme-catalyzed replication in modern biology with the very different nature of nonenzymatic RNA replication in simple protocells. In today's biology, whether we are looking at replication in humans, bacteria, or viruses, complex enzymes drive the process forward using the "energy molecule" adenosine triphosphate (ATP). A critical aspect of all enzymatic replication processes is that ATP-using enzymes called helicases separate the two strands of the duplex DNA, so that the single strands can be easily copied. Because of this procedure, replication can proceed at a constant temperature. It is not necessary to have periods of high temperature to separate the strands of a duplex, since the necessary energy can come from ATP. In contrast, with no enzymes and no ATP, replication during the origin of life had to proceed in a very different manner. What has been found experimentally is that temperature cycles are one way to accomplish seemingly contradictory requirements. For example, very short oligonucleotides (short nucleic acid polymers) can only bind to a longer template strand at low temperatures, meaning that a low-temperature environment is required for the extension of short primers. However, replication itself cannot proceed in a steady low-temperature environment, because longer oligonucleotides become "stuck" in stable duplexes; high temperatures are needed in order to separate the strands of such duplexes and allow copying to proceed. Consequently, an environment that cycles between high and low temperatures seems to be a requirement for nonenzymatic RNA replication. Moreover, RNA is a rather fragile molecule that degrades if exposed to heat for long periods of time. Therefore, the environment in which life began must have been cool most of the time, but with short spikes of high temperature to allow for strand separation. Remarkably, lakes in volcanically active areas (such as Yellowstone Lake) or asteroid impact craters (which typically contain crater lakes) provide for precisely such environments. Early cells could have been briefly immersed in the plumes of hot water that emerge from vents at the bottom of such ponds or lakes, only to be immediately cooled down once they hit the surrounding cold water of the lake. In other words, while evolved modern cells can survive in a stable, unchanging environment, primitive simple cells require a dynamic and constantly changing environment in order to drive the cyclic processes of reproduction.

Astonishingly, we now think that life on Earth might have emerged in a place very similar to Darwin's prescient "warm little pond." Based on experiments and observations of the types that we have described, this pond or lake had to be on Earth's surface, where it could have received UV radiation, and where the water was rich in certain minerals and metals, such as iron and phosphates. The pond also had to experience wet-dry cycles, during which the crucial ingredients could reach high concentrations. In our view, two locations that satisfy the criteria for being the birthplace of life on Earth are hot springs in volcanic areas, and craters created by asteroid impacts. Despite their differing origins, these environments have strong similarities, with hydrothermal circulation through cracked rocks bringing ions and chemicals to the surface in ponds or lakes, with the wet-dry cycles occurring on the banks of the ponds and temperature fluctuations occurring through vents releasing plumes of hot water. These similarities make it difficult to say which of these types of locations is more favorable as a cradle of life.

Once the first replicating protocells have established a foothold in some favorable local environment, and have begun to adapt to that environment through the process of Darwinian evolution, what might we expect to happen next? Presumably the earliest evolutionary advances would act to overcome the limitations of nonenzymatic

RNA replication and the need for environmentally triggered membrane growth and division. This could proceed through the evolution of multiple ribozymes acting to improve the efficiency and accuracy of these basic cellular processes. Evolving these ribozymes would require a gradually increasing capacity to maintain a larger genome. This, in turn, would lead to the evolution of new functions that would allow these still primitive cells to explore and adapt to new environments in which they could not initially have survived. The extent to which cells of the RNA World (i.e., prior to the evolution of coded protein synthesis) could spread into new environments is not at all clear. If an RNA-catalyzed metabolism was possible, then such cells might have been able to penetrate into progressively more nutrient-poor environments, as they evolved the ability to synthesize the missing nutrients internally. In any case it is clear that following the evolution of protein synthesis and thereby the appearance of structural proteins, membrane proteins, and enzymes, life began to spread across more and more of the planet. Today there is almost no conceivable environment that has not been colonized by life, including such challenging environments as deep-sea hydrothermal vents.

The ability of life to adapt, in an evolutionary sense, to new environments has important implications for the search for life elsewhere in our solar system. We know that there has been considerable exchange of materials between the planets that orbit our Sun. For example, many fragments of Mars that were ejected by impacts have landed on Earth as meteorites. Therefore, it is quite possible that microbial life has traveled between planets on the rocky ejecta of large impacts. As we begin to explore the rocky planets and the moons of the gas giants Jupiter and Saturn, the search for life is a high priority. But if we do find life, either extant life or signs of ancient extinct life, we must be prepared for the possibility that such life shares a common origin with life on Earth and is not the result of a truly independent origin of life. The search for signs of life on Mars is becoming more and more exciting as the technological capacity of our rovers increases. For example, NASA's *Perseverance* rover landed specifically in the Jezero crater on Mars, because evidence points to this crater having once been a lake that later dried out, thus potentially creating an embryonic habitat for emerging life. On the other hand, NASA's *Europa Clipper*, scheduled for launch in October 2024, will test the viability of subsurface oceans as environments for life. It will investigate Jupiter's icy moon Europa, precisely because this Jovian satellite shows strong indications of an ocean of liquid water underneath its icy crust. In Chapter 8, we'll explain why, in principle at least, this subsurface ocean could harbor conditions favorable for *sustaining* life. The key point is that even if life could not originate in vast oceans, oceans could still preserve and sustain life, once living organisms had been transported there. In the case of Europa, life could have been delivered via rocks ejected from, say, the surface of Mars, by powerful asteroid impacts.

Is life common or rare in our galaxy and the universe beyond? The conclusion is clear. It is virtually impossible to rigorously and comprehensively understand the origin of life when all we have at our disposal is just one example of life. This reality forces us to search for signs of life, or at least for signatures of prebiotic chemistry, beyond the boundaries of our home planet.

# CHAPTER 7

# Extraterrestrial Life on Solar System Planets?

*Are we such apostles of mercy as to complain if the Martians warred in the same spirit?*
—H. G. Wells, *The War of the Worlds*

Naturally, the first place to search for extraterrestrial life would be on other objects in the solar system. It is much easier to study, either remotely or directly, the conditions on planets revolving around the Sun, or even on the moons orbiting those planets, than to examine in detail extrasolar planets. The reason is simple—solar system objects are just so much closer. The relative proximity allows us to explore solar system planets better not only with telescopes, but also with probes that can either fly by these "wanderers" of the night sky (as the ancient Greeks referred to planets), orbit them, or even (in some cases) land on their surface and conduct in situ observations and experiments.

Of all the solar system planets (other than Earth), there is no doubt that humans have always found Mars to be the most alluring and most likely to harbor life. The fascination with the red planet has also inspired numerous science fiction books. These range from *Two*

*Planets* (published in 1897) by the "father of German science fiction," Kurd Lasswitz, and the famous and influential *The War of the Worlds* by the prolific author H. G. Wells (1898), to *A Princess of Mars* (1912) by Edgar Rice Burroughs (creator of *Tarzan*), and the short-story collection *The Martian Chronicles* (1950) by Ray Bradbury (of *Fahrenheit 451* fame). These and other books told stories of a Martian society living in a breathtaking world, and engaging in hair-raising adventures, which included exploration of Earth's North Pole, fierce battles, brilliant swordplay, and violent invasions aimed at getting hold of mineral resources. The latter stories became part of the rich invasion literature in England at the end of the nineteenth century. These imaginary tales were undoubtedly inspired on the one hand by Mars's similarity to Earth (just as an example, a Martian day is longer than a day on Earth by only about forty minutes), and on the other, by the erroneous belief in the late nineteenth century that there was an elaborate network of "canals" on the surface of Mars. Those fictitious canals turned out to be nothing but straight-line optical illusions produced by the low-resolution telescopes of the time. The linear artifacts were first observed in 1858 by the Italian Jesuit priest and astronomer Angelo Secchi, and then mapped in 1877 by Italian astronomer Giovanni Schiaparelli. Both astronomers described the features in Italian as *canali* (meaning "channels" and referring to natural configurations)—a word that was mistranslated into "canals"—giving the impression that they represented an advanced irrigation system designed and excavated by intelligent beings. Somewhat inexplicably, American businessman, author, and astronomer Percival Lowell became a fervent advocate of the "canals" interpretation, and he spent much of his career attempting to prove that Mars was inhabited by an intelligent civilization—a concept enthusiastically embraced for about half a century by many in the general public. In fact, both of us, the authors, remember that during our elementary school days, quite a few people believed that Mars could harbor an intelligent civilization.

Fascinatingly, the co-originator of the theory of natural selection in the evolution of life on Earth, Alfred Russel Wallace, published in 1907 a book in which he presented an insightful and scathing criticism of Lowell's ideas, concluding with the uncompromising words: "Mars, therefore, is not only uninhabited by intelligent beings such as Mr. Lowell postulates, but is absolutely UNINHABITABLE" [capitalized in the original].

Astronomers had indeed become increasingly skeptical about the existence of a Martian civilization as the twentieth century progressed. Still, back at the end of the nineteenth century, a sufficient number of astronomers were so convinced that Mars was inhabited, that when the "Pierre Guzman Prize" of 100,000 francs was established in France, to be awarded to the first person to communicate with other planets, communication with Mars was specifically excluded, because that was considered far too easy!

Disappointingly for many scientists, and indeed the general public, all of those romantic fantasies about Mars were rapidly quenched as soon as the age of space exploration began. In particular, the images of the Martian surface returned by the *Mariner 4* spacecraft in 1965, and later images from its more advanced successor missions—the 2008 *Phoenix* lander and the 2012 *Curiosity* rover—revealed that Mars was nothing but a frigid, arid, empty desert, pockmarked with craters like the Moon, and possessing a very thin, extremely low-pressure atmosphere. Worse yet, dispirited scientists discovered that Mars did not even have a global magnetic field (only relatively small patches of magnetized crust). In other words, Mars lacked two of Earth's main defense systems. Earth's atmosphere on one hand keeps Earth's mean surface temperate at about sixty degrees Fahrenheit (fourteen degrees Celsius), and on the other, it protects life on Earth from harmful ultraviolet radiation from the Sun, and from energetic cosmic rays from deep space. Similarly, Earth's global magnetic field shields Earth's atmosphere from erosion by the solar wind (charged particles that the Sun continually spews), by deflecting those particles away from our planet.

Even with these sobering realizations, however, Mars remains the main target in the search for extraterrestrial life in the solar system, albeit with greatly revised expectations. Those were basically reduced to the mere hope of finding evidence for past (or present) microbial life.

## Mars Is There, Waiting to Be Reached

Buzz Aldrin walked on the Moon in 1969 as part of the Apollo 11 mission. Since then, he has been continuously expressing the ambition encapsulated by the title of this section. Like mountaineer George Mallory before him, who gave the famous answer "because it's there" as his reason for wanting to climb Mount Everest, Aldrin created another motivational phrase, which by now is almost invariably trotted out whenever someone tries to justify a not-so-clearly justifiable ambition. Entrepreneur and adventure seeker Elon Musk went even further. He declared that he wants to die on Mars, though "not on impact." This may be an enticing goal to some, as evidenced, for example, by the immense success of the 2015 sci-fi movie *The Martian*, but it has relatively little to do with the scientific search for life. While there's no denying that NASA's *Perseverance* rover, which has been traveling across the twenty-eight-mile-wide Jezero crater on Mars since February 2021, may miss some startling discoveries that human geologists would not have overlooked, the fact is that machine learning, and artificial intelligence in general, are advancing at a meteoric pace, as is sensor technology. Consequently, there is little doubt that exploration by smart robots is in the long run the safer and cheaper way to proceed.

There are still strong reasons to believe that there could be (or could have been in the past) some form of life on Mars. One such reason is the fact that about 4.5 billion years ago, Earth and Mars formed from roughly the same patch of the protoplanetary disk of gas and dust that surrounded the young Sun. This means that Mars was *at least initially* endowed with similar ingredients to those that proved to be sufficient for life to have emerged on Earth. Those included water, which could act as a solvent, bringing the relevant molecules into contact and allowing numerous reaction paths. Very basic building blocks of organic material, in the form of molecules of hydrogen, ammonia, and carbon monoxide, were also likely to have been abundant on the early Mars. But it isn't just the availability of some of the fundamental components of life that gives reasons for optimism when it comes to assessing the probability for a past life on Mars. There are also many lines of evidence suggesting that the initial Martian environment was much more auspicious to life than it is today. First, images taken by a number of orbiting satellites have shown unambiguous traces of flowing water, in the form of rivers, streams, and lakes of various sizes. The geography and topography of the Jezero crater itself, where the *Perseverance* rover has been exploring, also indicates that, more than three billion years ago, the crater was a body of water about the size of Lake Tahoe, with rivers flowing into it. The presence of large boulders, washed into Jezero, points to episodes of massive flooding. This means that liquid water existed on the early Martian surface, probably for at least millions of years. In turn, the evidence for past liquid water indicates that the Martian atmosphere was at that time sufficiently thick, capable of insulating the planet and keeping it considerably warmer than its current frigid global mean surface temperature of about minus eighty degrees Fahrenheit (minus sixty-two degrees Celsius). A thicker atmosphere could also have protected the surface from destructive short-wavelength UV radiation, which would have broken down the complex molecules needed to originate life. The hunt for liquid water on Mars, by the way, is a constant and continuing theme in the investigations of the red planet. One of its latest finds, announced by the European Space Agency on December 15, 2021, was that their *ExoMars Trace Gas Orbiter* mission had discovered "significant amounts" of water hidden underneath the surface of the Valles Marineris canyon system, an area about ten times longer and five times deeper than Earth's Grand Canyon. The detection by China's *Zhurong* rover (announced in 2023), of irregular polygonal wedges buried about 115 feet under the Martian surface, also may imply the existence of freeze-thaw cycles on Mars some 2.9 to 3.7 billion years ago, providing further support for the presence of liquid water on ancient Mars.

Given that there are signatures of biological activity on Earth as early as some 3.5 billion years ago, and considering the fact that Mars enjoyed quite similar initial conditions, it is conceivable that around the same time, that is, almost four billion years ago, comparable prebiotic chemistry could have started to operate on Mars. This is especially likely since, as we have noted in Chapter 6, lakes such as the one that existed in the Jezero crater, rather than large oceans, were most likely the places where life emerged on Earth. Consequently, microbial life-forms, perhaps not too different from those that appeared on Earth, or at least some prebiotic signatures, could in principle be found on (or buried in) the Martian surface. Such potential discoveries would provide invaluable clues concerning the environments that led to the birth of life on Earth—environments that are inaccessible on Earth due to the history-deleting action of plate tectonics.

But even if life did emerge on Mars, it clearly never came even close to becoming as complex as life on Earth. Why have the two planets diverged so much in their evolutionary paths? To attempt to understand why Mars is such a desolate world while on Earth there are no fewer than about nine million species, we have to examine in more detail the differences, rather than the similarities, between the two planets.

The contrast between Earth and Mars results primarily (directly or indirectly) from a few physical properties. In particular, the size of the planet—the diameter of Mars is roughly half of that of Earth—and the planet's surface gravity—that of Mars is almost three times weaker than that of Earth (because its mass is only slightly more than one-tenth of Earth's mass). Smaller planets cool faster because the ratio of their surface area to their volume is larger, and therefore they are prone to losing heat at a higher rate. Earth and Mars were initially both molten due to the combined heating actions of gravitational accretion of mass when the planets formed, asteroid impacts on the surface, and radioactive decays (for example, isotopes of uranium, thorium, and potassium) in the interior. Consequently, their early internal structures were probably quite similar, consisting of a molten metal core, surrounded by rocky outer layers. As we have mentioned earlier, there is also evidence suggesting that the two planets had at early times a relatively thick atmosphere (probably fed by active volcanoes). This is, however, where the similarities end. Because of Mars's weaker gravity (something weighing one hundred pounds on Earth weighs only thirty-eight pounds on Mars) and the lack of a global magnetic field, energetic particles from the solar wind managed to strip Mars of most of its atmosphere, leaving behind only a whiff of carbon dioxide by the time the planet was just about one billion years old.

From the perspective of being able to allow for life to evolve, the consequences were devastating. We know, for instance, that the boiling temperature of water drops as the atmospheric pressure decreases. On Mars, whose atmospheric pressure is currently about 160 times lower than that on Earth, there was no way for liquid water to be sustained on the surface. Thus the loss of the atmosphere was inevitably followed by the loss of the planet's surface water. Minerals may offer clues as to how the red planet's climate changed from being more Earth-like to that of the frozen desert it is today. This is why NASA's *Curiosity* rover arrived in the fall of 2022 to the "sulfate-bearing unit," a region enriched with salty minerals where the rover also found evidence of an ancient lake in the form of rocks etched with the ripples of waves. Scientists hypothesize that billions of years ago, streams and ponds left those minerals behind, as the water dried up. Clay minerals, and in particular the minerals known as *phyllosilicates*, are some of the most interesting minerals discovered on Mars, due to their roles as indicators of water-rock interaction. Their types, sites, and abundances provide clues to ancient environmental conditions of Mars and to the potential locations where mineral-bound water and perhaps biosignatures are likely to be detected today.

Another feature that might have been consequential for the development of complex life, or lack thereof, is plate tectonics. In the 1950s, geologists started to realize that plate motion, and the ensuing continuous recycling of the crust, can produce earthquakes and volcanic activity—most of Earth's volcanoes are found at the border of tectonic plates. More relevant for our topic, however, a wave of research that started around 2015 suggests that plate tectonics may have also been crucially important for the emergence and evolution of life (as discussed below). If true, this could be an additional reason as to why Earth is so vibrant with life, while Mars appears to be bereft of life.

What the new research about the potential role of plate tectonics in the evolution of life submits is that, among other things, plate tectonics ensured the longevity of Earth's atmosphere, and its favorable composition for life. For example, plate tectonics probably played a role in regulating the concentration of carbon dioxide in Earth's atmosphere, and thereby (since carbon dioxide traps surface heat through the greenhouse effect), it may have operated as Earth's "thermostat" on long timescales. Put simply, here is how this scenario could have worked. Carbon dioxide in the atmosphere dissolves in rainwater and the resulting acidic mixture erodes rocks and flows into the oceans. Calcium from rock minerals combines with the dissolved carbon dioxide, creating limestone at the ocean floor. Via plate tectonics, the limestone and other minerals are continuously carried to subduction zones, melted there, and carbon dioxide is released back into the atmosphere via volcanic eruptions. In addition, and again through the process of subduction, plate tectonics causes even the water in the oceans to be cycled through Earth's mantle, thereby renewing the ocean floor.

For decades, planetary scientists believed that Mars is geologically dead. Its crust was assumed to be composed of one gigantic plate. Such planets are known as "stagnant lid" planets. While NASA's *InSight* spacecraft (which touched down on Mars in 2018) had recorded by 2022 more than a thousand distinct "Marsquakes," those were generally thought to be caused simply by rock fractures in the Martian crust due to shrinkage of the planet as it cools down.

Most geologists have long agreed that even if Mars had some tectonic activity in its distant past, it hasn't had any for at least three billion years. As geophysicist Bradford Foley of Pennsylvania State University puts it: "There is some argument that maybe very, very early on, it could have had plate tectonics, but my view is it probably never did." In a paper published in December 2022, however, planetary geophysicists Adrien Broquet and Jeffrey Andrews-Hanna of the Lunar and Planetary Laboratory at the University of Arizona argued that data from a few robotic missions show that a mantle plume—a 2,500-mile-diameter column of hot material moving upward from underneath the crust of the Elysium Planitia region on Mars—is creating tectonic activity, demonstrating that Mars is still geodynamically active today. In principle, such a hot mantle plume could cause at least some of the water below Mars's surface to be in liquid form (supporting perhaps the possibility of a subsurface biosphere).

At the same time, we should note that Foley himself does not think that plate tectonics are indeed a necessary requirement for sustaining life, at least not on Earth-size planets. In work he published in 2018 with geoscientist colleague Andrew Smye, the researchers used computer models to show that even stagnant lid (Earth-size) planets can maintain carbon dioxide levels (through volcanic outgassing) suitable for habitability for a few billion years. The only ingredients that were needed were sufficient internal heating through radioactive decays, and an initial carbon dioxide budget that was not lower than 1 percent of that of Earth. Foley and Smye therefore concluded that the initial composition and size of the planet are the most important characteristics in setting the trajectory for habitability.

All of these and similar studies inevitably lead us to the following question: *Have we ever detected any signs of past or present life on Mars?* Oddly, the answer to this seemingly unambiguous question depends somewhat on who you ask. Mars, as it turns out, has provided us over the years with a series of heated contentions and unsolved mysteries.

## Mars Then Belongs to the Martians

The first controversial result concerning life on Mars came almost five decades ago. In 1975, after a frantic race against the clock, NASA managed in the nick of time to put two biological experiments on board two identical spacecraft, *Viking 1* and *Viking 2*, just before they were launched (three weeks apart) on their way to Mars. The first lander touched down on July 20, 1976, in Chryse Planitia ("Golden Plain"), a flat lowland region in Mars's northern hemisphere, and the second one landed on September 3, 1976, in Utopia Planitia ("Nowhere Land Plain"), a lava plain some four thousand miles away. Among other tasks, these landers performed experiments that searched for signs of microbial life in the Martian soil. Unexpectedly, the life-detection experiments on both landers produced the same extremely contentious results. One analysis in particular, dubbed the *Labeled Release* (*LR*) experiment, was based on the widely accepted notion that metabolism constitutes a universal characteristic of life. The test therefore mixed Martian soil with a nutrient that contained radioactive carbon. The expectation was that any microbial life (if it was present) would produce radioactive gases through the chemical reactions of metabolism. Amazingly, the LR on board *Viking 1* indeed showed right away that the Martian soil tested positive for metabolism—a sign that on Earth would have been taken to suggest the presence of life. However, a second, parallel experiment on the same spacecraft found no trace of organic material in the soil, suggesting the complete absence of any organics-based life. A second addition of nutrients to the soil sample did not result in additional release of labeled carbon dioxide, which is more consistent with an oxidizer in the soil than with life. The experiments on board the *Viking 2* lander gave very similar results. The most important lesson to be drawn from all of this is that we have to expect surprises, and the surprising chemistry of Mars's soil is a great example of how easy it is to be fooled by false positive signals that look like life but really aren't.

Even though most researchers attributed the positive LR results to the presence of some unidentified non-biological oxidants in the Martian soil, in the nearly fifty years that have passed since the analyses were performed, scientists have still been unable to fully and univocally reconcile the conflicting results. Whereas the general consensus was, and still is, that the *Viking* landers found no convincing evidence for life on Mars, a small minority of scientists continues to argue that the *Viking* results were positive for life. In 2012, an international team of scientists led by biologist Giorgio Bianciardi of the University of Siena analyzed the results of the LR experiment using an exploratory data analysis technique known as "cluster analysis." The team concluded: "These analyses support the interpretation that the Viking LR experiment did detect extant microbial life on Mars." Gilbert Levin, who was the principal investigator for the LR experiment (and a member of the 2012 international team), and biochemist Patricia Ann Straat, who was also a member of the LR team, remained quite convinced that their experiment had discovered life on Mars. In a paper they published as late as 2016, they concluded that "extant life is a strong possibility, that abiotic interpretations of the LR data are not conclusive, and that . . . biology should still be considered as an explanation for the LR experiment." In an opinion piece Levin posted in 2019 in *Scientific American*, he doubled down on this conclusion, saying: "A panel of expert scientists should review all pertinent data of the *Viking* LR together with other and more recent evidence concerning life on Mars. Such an objective jury might conclude, as I did, that the *Viking* LR did find life." Sadly, Straat passed away in 2020 and Levin passed away in 2021.

Surprisingly, in 2023, astrobiologist Dirk Schulze-Makuch, from the Technical University of Berlin, speculated that NASA may have in fact discovered life on Mars when it first put its two *Viking* landers on the red planet, but that the agency may have also accidentally killed it by drowning, when the experiment added water to the soil. His speculation was inspired by the existence of microbes living inside salt rocks in the Atacama Desert in Chile that do not need rain to survive and that could be eradicated by too much water. We should be clear though: the *Viking* experiments were generally seen as having given negative results for life, and that spelled disappointment and frustration for most planetary scientists. Perhaps none was more disheartened than Carl Sagan, who still diligently, but unsuccessfully, attempted to find some signs of life in the many images taken by the *Viking* spacecraft from orbit. Nevertheless, in his celebrated book *Cosmos*, Sagan appeared to not have completely given up on life on Mars: "If there is life on Mars, I believe we should do nothing with Mars. Mars then belongs to the Martians, even if the Martians are only microbes. The existence of an independent biology on a nearby planet is a treasure beyond assessing, and the preservation of that life must, I think, supersede any other possible use of Mars."

## The Rock

Remarkably, the *Viking* results were not the only controversial findings related to the possibility of life on Mars. Another presumed discovery resulted in no less than a special presidential proclamation on August 7, 1996. On that date, President Bill Clinton announced to the world from the South Lawn of the White House: "Today, rock 84001 speaks to us across all those billions of years and millions of miles. It speaks of the possibility of life. If this discovery is confirmed, it will surely be one of the most stunning insights into our universe that science has ever uncovered. Its implications are as far reaching and awe-inspiring as can be imagined."

"Rock 84001" referred to a meteorite discovered in Antarctica a dozen years earlier by Roberta Score, a meteorite expert with the United States Antarctic Program. It was labeled ALH84001, because it was the first meteorite found at Allan Hills in 1984 (hence the designation 84001). NASA geologist and astrobiologist David McKay and his team received the 4.3-pound (1.93 kilograms) rock to be analyzed in the early 1990s, and the first findings were already quite exciting. By examining how long the rock had been exposed to cosmic rays, its age was determined to be not that much younger than the age of the solar system itself—about 4.1 billion years. This in itself made the meteorite interesting, since it was the only Mars meteorite originating at a time when Mars may have still had liquid water on its surface. Subsequent studies of the rock's composition, and the discovery and analysis of gas trapped within it (which was found to have an identical composition to that of the Martian atmosphere), revealed that the rock had most likely been launched from the surface of Mars by a powerful impact that occurred about seventeen million years ago, and it landed in Antarctica some thirteen thousand years ago. The fact that the meteorite arrived to Earth as a "visitor" from Mars was not in itself a huge surprise—more than three hundred meteorites have been classified as being of Martian origin. The largest intact Martian meteorite to date, Taoudenni 002, was recovered in Mali in early 2021. In fact, NASA scientists colloquially say that Mars and Earth have been "swapping spit" for billions of years. In other words, when either planet is hit by asteroids or comets, some ejecta shoot into space, and a small fraction of that material can land on the other planet. To McKay's amazement, however, ALH84001 showed several unique features. First of those was the presence within the rock of certain organic compounds known as PAHs (*polycyclic aromatic hydrocarbons*), which, while quite common on Earth and in other places in the solar system, were similar to the type that usually accompanies the decay or burning of organic matter. A second surprise was the simultaneous existence of three compounds:

both rounded brownish and clear carbonate globules, iron sulfide minerals, and magnetic minerals consisting of an oxide of iron—a mixture that is rarely produced all together through non-biological processes, but which can be concurrently synthesized by some bacteria. Finally and most bewildering, McKay discovered that some of the observed structures within those globules bore an uncanny resemblance in shape to fossilized Earth bacteria, albeit tiny ones, only a few tens of nanometers (in total about a millionth of an inch) across. At that point McKay became quite convinced that he and his team had discovered the first evidence for ancient extraterrestrial life, and once a few independent, professional referees for the journal *Science* accepted McKay's paper on the findings for publication, a NASA press conference and President Clinton's declaration became almost unavoidable.

Unfortunately, the echoes of the press conference had barely died away when criticisms started pouring in from all directions. By the time everything was said and done, other researchers (including McKay's brother Gordon!) had shown that the data presented by McKay's team to support the case for discovering life in the meteorite, including the carbonate globules, the PAHs, the magnetic crystals, and the "nanobacteria" (the tiny structures resembling life-forms), could all be explained as resulting from abiotic chemical processes. In particular, the nanobacteria were shown to represent mere transformations of minerals from being amorphous to structures mimicking life, through the process of crystallization. The consensus of the scientific community was, therefore, that in their interpretation of the ALH84001 results, McKay's team members fell victim to relying too much on morphology—the shapes of the structures—usually considered a poor indicator of something having been produced by living organisms.

In one of his last essays before prematurely passing away, even Carl Sagan admitted that "the evidence for life on Mars is not yet extraordinary enough." A 2012 review of the results from the ALH84001 meteorite concluded on the basis of all the data available by that time: "This scenario evokes the biological equivalent of an Occam's razor whereby it is easier to accept the carbonate globules and bacteria-like structures seen in ALH84001 as the result of universally prevalent chemical reactions than to accept them as proven signs of extraterrestrial life." This conclusion was further strengthened by additional research published in January 2022, in which biochemist Andrew Steele from the Carnegie Institution for Science in Washington, DC, and collaborators showed that the organic matter in the rock probably formed from the chemical interplay of water and minerals mixing under Mars's surface.

The *Viking* and ALH84001 stories bring forth two interesting lessons. First, when one deals with discoveries of this magnitude, a healthy amount of skepticism is an excellent practice. Again, a maxim that Sagan coined early on in his career can serve as a good guide: "Extraordinary claims require extraordinary evidence." Second, the findings uncovered features that can mimic some of the characteristics of life as we know it: metabolism in the case of *Viking*, morphology in the case of the meteorite. This second lesson should caution against "false positives"—potential discoveries that do not truly represent life. We should be mindful of not being too biased in favor of life as we know it.

In any case, as we have noted already several times, identifying even abiotic sources of organic compounds that could have operated on the early Mars may aid in discerning the types of conditions that could lead (or not) to the emergence of life. The simple reason is that while the signatures of prebiotic chemistry on Earth have not been well preserved (because of plate tectonics), on Mars, rocks older than 3.5 billion years are quite common.

## Now You See It, Now You Don't

Mars has presented planetary scientists with yet another mystery, this time related to the presence (or not?) of methane. Most people are familiar with the fact that on Earth a non-negligible amount of methane is produced by the microbes that help digestion in livestock, with flatulent cows and sheep ending up burping the gas into Earth's atmosphere. It is also produced by termites (through their digestive processes), by volcanoes, and by deposits underneath Antarctic ice and Arctic permafrost.

Methane, which is the chief constituent of natural gas, is a chemical compound composed of one carbon atom bonded to four hydrogen atoms. Methane can be formed through abiotic reactions, such as the reduction of carbon dioxide by the hydrogen generated in complex rock-water reactions called serpentinization. Alternatively, methane can also be generated by the bacterial fermentation of organic matter, and on Earth this contributes significantly to global warming. Biologically derived methane is generated on a large scale both in the thawing permafrost of Arctic regions, and in the digestive tracts of ruminants such as cows and goats.

We know that there are no cows or goats on Mars, but since methane can be produced by living microorganisms, it has always been considered (in combination with other gases) a potential biosignature (a sign of life), and its detection has been a coveted goal for those searching for life on Mars (and indeed, as we shall see in Chapter 9, on exoplanets as well). However, we should always remember that methane can also be created by a variety of non-biological processes, including processes that could happen on Mars. The latter include reactions between the igneous iron-magnesium silicate mineral known as olivine, which is abundant on Mars, and water and carbon dioxide, in high-pressure, high-temperature subsurface environments.

Here, however, comes the mystery. Since 2003, a few instruments have detected methane on Mars, while others, which were equally sensitive, have not. As a specific, very intriguing example, NASA's *Curiosity* rover carried a sophisticated suite of instruments collectively known as *SAM* (for *Sample Analysis at Mars*), which repeatedly detected methane in the Martian atmosphere as the rover was speeding over the surface of the Gale crater. On average, SAM measured about 0.4 parts per billion of volume—the equivalent of a quarter of a teaspoon of sugar dissolved in an Olympic-size pool. By June 7, 2018, SAM had detected even seasonal variations in the methane levels, as well as puzzling spikes towering almost fifty times above the average value. At the same time, however, another instrument, designed to be the gold standard for measuring methane concentrations—the *Trace Gas Orbiter* (*TGO*) on board the European Space Agency's *ExoMars* (launched in March 2016)—failed to detect any traces of methane higher up in the Martian atmosphere. The *TGO* recorded nothing but null detections (with upper limits as low as 0.02 parts per billion of volume) up to May 2019. To complicate things even further, researchers analyzing data from a spectrometer—a device that can determine composition of a substance by the analysis of the spectrum of emitted light—on board the European Space Agency's *Mars Express* orbiter (launched in 2003) reported in 2019 that it had recorded a spike in the methane concentration in the Martian atmosphere above the Gale crater. Moreover, that data had been collected on June 16, 2013, just one day after the in situ observation of a methane spike by the *Curiosity* rover.

This situation created a baffling conundrum, since it forced researchers to first attempt to solve the puzzle of why certain instruments had made positive detections while others had not, before they could even turn to the more interesting and potentially consequential question of what could be the source of the methane. Chris Webster of NASA's Jet Propulsion Laboratory in California, the lead scientist of the spectrometer in the SAM laboratory, could hardly hide his amazement when presented with the conflicting results: "When the European team announced that it saw no methane, I was definitely shocked," he admitted. Still, Webster immediately knew what he had to do. He and his team promptly engaged in examining all the SAM measurements, to check for the remote possibility that the *Curiosity* rover itself was somehow releasing the gas. They inspected the data for correlations between detections and a series of rover conditions, such as the direction in which the rover was pointing, how its wheels were turning, or whether it was crushing rocks. They found nothing.

At this point, Canadian planetary scientist John Moores of York University came up with an unexpected, provocative question, which sounded almost as if it had been taken straight out of an old Jewish joke. The joke goes like this:

> *Two neighbors were fighting over a financial dispute. They couldn't reach an agreement, so they took their case to the local rabbi. The rabbi heard the first litigant's case, nodded his head, and said, "You're right."*
>
> *The second litigant then stated his case. The rabbi heard him out, nodded again, and said, "You're also right." The rabbi's attendant, who had been standing by, was justifiably confused. "But, rebbe," he asked, "how can they both be right?" The rabbi thought about this for a moment before responding, "You're right, too!"*

The surprising question that Moores asked himself was, *Could both* Curiosity *and the* Trace Gas Orbiter *be right?* For that to happen, Moores and colleagues suggested that the discrepancy was simply the result of the time of day at which the measurements were taken. Specifically, because SAM's spectrometer required considerable power, its measurements were taken primarily in the Martian night, when other *Curiosity* instruments were not operating. At night, it turns out, the Martian atmosphere is usually calm, which would have allowed methane seeping from the ground to accumulate close to the surface, making it potentially detectable by SAM. On the other hand, the *Trace Gas Orbiter* operated during the day, when air circulation could have diluted the methane by incorporating it into a much larger air mass. To test this bold hypothesis, the *Curiosity* team led by Paul Mahaffy, the principal investigator of SAM, made a few experiments in which they bracketed a nighttime measurement by two daytime ones. The results were generally in agreement with Moores and his colleagues' prediction—the two daytime measurements gave no detections, while the nighttime one was consistent with *Curiosity*'s earlier measurements, seemingly corroborating the idea that the concentration of the methane near the surface of the Gale crater changes during the day. As it turned out, however, even these promising results did not provide for an ironclad explanation. More detailed Martian weather simulations, by researcher Daniel Viúdez-Moreiras of the National Institute for Aerospace Technology in Madrid, Spain, and collaborators, have shown that a mystery still remains, and maybe even thickens. The simulations indicated that very small methane emissions coming from the northwestern rim of the Gale crater (very close to the location of the *Curiosity* rover), *and only from that area*, could indeed lead to a detection by *Curiosity* and a non-detection by the *TGO*, but that this solution in itself is rather improbable or problematic. Basically, what is required is one of two possibilities: either that there exists a strong and unknown loss mechanism in the atmosphere that prevents the accumulation of global methane (since the estimated lifetime of methane in the Martian atmosphere is about three hundred years), or that methane emissions are extremely rare on Mars and the *Curiosity* rover had fortuitously landed next to one of them. Otherwise, one would have expected the *TGO* to still detect methane in the entire Martian atmosphere.

Concerning the potential source of the putative methane, a few studies have argued that the methane spikes could be coming from microorganisms that produce methane (*methanogens*) forming a biosphere a few miles underneath the Martian surface, where liquid water may still exist. Nevertheless, while these studies have shown that the observed abundance of atmospheric methane is at least consistent with the existence of such a biosphere, they should definitely not be taken as *proof* for any form of life on Mars.

Another interesting result concerning possible ancient life on Mars came from another crater. In 2007, the Mars Exploration Rover *Spirit* encountered rocks and regolith (dust and broken rocks) composed of hydrated silicon dioxide (known as opaline silica) next to a volcanic landform in Mars's Gusev crater. Such silica deposits have long been targets in the search for fossil life on both Mars and early Earth, because of their ability to capture and preserve biosignatures. By comparing this silica to that found at the Roosevelt Hot Springs and Opal Mound in Utah, and also to that at the El Tatio geyser field and the Puchuldiza-Tuja hydrothermal system in Chile, planetary geologist Steven Ruff of Arizona State University and collaborators concluded in 2020 that the Mars opaline silica rocks are deposits from hot spring/geyser activity and that the morphology of the silica structures bears a strong resemblance to the microbially mediated microstromatolites at El Tatio. This is significant (even though it certainly requires much further study) because, as we explained in Chapter 6, environments with hot springs are considered to be promising targets in the search for ancient life. On Earth, for example, opaline silica preserved evidence of microbial life throughout geological history, and fossilized hydrothermal fields extend as far back as the 3.48 billion-year-old deposits of the Dresser Formation in the Pilbara Craton of Western Australia.

The conclusion is that, somewhat disappointingly, even though there have been a few tantalizing hints (albeit ambiguous and controversial), there is no convincing evidence for life on Mars. Still, the negative results in the search for life on Mars so far do not mean that the study of Mars will not or should not continue. The lack of detections definitely does not indicate that life never existed on Mars. The *Perseverance* rover arrived to a dried-up river delta on the west rim of the Jezero crater at the end of May 2022. By May 28, *Perseverance* had ground a two-inch-wide circular patch into a rock at the base of the delta to collect a sample. Overall, *Perseverance* brought to Mars forty-three tubes, of which thirty-eight were for collecting samples. By the end of October

2023, the rover had already collected twenty-three samples from rocks, regolith, and the atmosphere. The delta in the crater formed billions of years ago, when that long-vanished river deposited layers of sediment as it entered the crater lake. On Earth, river sediment is usually teeming with life, so researchers hope that the Mars grains will also contain chemical or other traces of past life. These hopes were reinforced by the fact that ground-penetrating radar on board *Perseverance* confirmed that the Jezero crater, formed by an ancient meteor impact, once harbored a lake and a river delta. As Imperial College London planetary geologist Sanjeev Gupta puts it: "Poet William Blake's 'To see a world in a grain of sand' comes to mind." NASA and the European Space Agency plan to retrieve those *Perseverance* samples and fly them back to Earth for detailed study. This will happen no earlier than 2033, and it would be the first-ever sample return from Mars. Significantly, NASA is already planning the construction of a facility that will house these samples, since even though the chances of a "Martian pandemic" are very low, the structure should be capable of safely containing potentially dangerous Martian pathogens. At the same time, the facility also has to be pristine, to prevent substances on Earth from contaminating the samples from Mars. Understandably, no one knows what to expect. Kenneth A. Farley, a geochemist at the California Institute of Technology who is the project scientist for the *Perseverance* mission, is not willing to predict what they will find, merely noting: "Let's just say we are not going to bet."

Missions to Mars will undoubtedly continue, and a few of them will eventually even involve human crews, but there are other objects in the solar system that also deserve our attention.

## The Shining Goddess of Love

Before the space age one might have expected that Earth's sibling planet—Venus—which is roughly the same size and mass as Earth, and which is constantly covered with clouds, would be lush with jungles and muddy swamps. Modern observations by the Soviet Union's *Venera* missions, and by NASA, the European Space Agency, and Japanese spacecraft, have radically changed our views on this celestial object. In spite of its beauty—being, after the Moon, the brightest object in the night sky—Venus turned out to be a hellish place that would make Dante's *Inferno* look like Paradise. Venus's toxic atmosphere is dominated by carbon dioxide, with nitrogen, sulfur dioxide, carbon monoxide, and water vapor, among the other volatiles (compounds that can be readily vaporized). As if this harsh composition wasn't enough, it is accompanied by a layer of clouds composed of droplets of sulfuric acid. The clouds trap heat like a greenhouse, making Venus so hot, with a surface temperature of almost 900 degrees Fahrenheit (480 degrees Celsius), that lead would melt on its surface. In addition, the atmospheric pressure at Venus's surface is a crushing ninety-fold that of Earth. In short, it is about as not-Earth-like as you can imagine. Consequently, for decades, Venus had completely dropped off the list of potentially habitable objects in the solar system, and as a result, the planet is currently orbited by a single probe—the Japanese spacecraft *Akatsuki*. This "abandonment" is about to dramatically change. On June 2, 2021, NASA approved not just one, but two new missions to Venus, and the European Space Agency (ESA) also approved a mission only a week later. NASA's *VERITAS* mission (short for Venus Emissivity, Radio science, InSAR, Topography, And Spectroscopy) consists of an orbiter that will image and map Venus's surface, with the goal of studying the planet's geological past. ESA's *EnVision* mission is an orbiter as well, which will use radar to map the surface, a sounder to reveal underground layering, and spectrometers to study both the atmosphere and the surface. NASA's second mission, *DAVINCI*, will include both an orbiter and a descent probe to Venus. The orbiter will image Venus in multiple wavelengths from above, while the descent probe will study the chemical composition of Venus's atmosphere and take photographs during its plunge.

In addition to addressing the broader perplexing puzzle of why the evolutions of Earth and of Venus have been so different that Venus turned into a sulfurous inferno, the three new missions (to be launched in the late 2020s and early 2030s) will attempt to answer a few specific questions, all of which are related to the potential for habitability either in the distant past, or, remarkably, even now. First and foremost, by measuring the precise gas composition of Venus's atmosphere, astronomers will strive to determine whether the early Venus had bodies of liquid water on its surface, which later boiled off because of a runaway greenhouse effect. Second, astronomers will use the high-resolution surface mapping that will be performed by *VERITAS*, and especially by *EnVision*, to find out whether there are active volcanos on Venus, since, as we have seen, geothermally active regions could have been the birthplace of life on Earth. Third, and related to the potential past existence of oceans, the probes will help scientists to find out whether there were continental landmasses on Venus's surface. In particular, *VERITAS* and *DAVINCI* will examine Venus's tesserae—regions of seriously deformed terrain characterized by high topography—which cover about 7.3 percent of Venus's surface, and which could represent something similar to continents. These future missions will also be able to test specific models of the evolution of Venus's atmosphere. For example, in a paper published in October 2023, planetary scientist Matthew Weller and colleagues concluded based on computer simulations that in order to match its present composition, the Venusian atmosphere had to be fed by volcanic outgassing in an early phase of plate-tectonic-like activity. Weller's findings suggest that Venus's atmosphere had to pass through a great climatic-tectonic transition, from an early phase of mobile-lid tectonics that lasted for at least a billion years, to the current stagnant-lid-like mode of reduced outgassing rates and essentially no horizontal tectonics compared with Earth's plate tectonics.

Renewed interest in Venus's candidacy as a habitat for life was sparked unexpectedly in 2020, when a group of astronomers, led by Jane Greaves of Cardiff University in Wales, tentatively discovered an interesting chemical—*phosphine*—high in Venus's cloud decks. Greaves later explained that although she had thought that the chances of finding phosphine on Venus were slim to none, she was "intrigued by the idea of looking for phosphine, because phosphorus might be a bit of a sort of go-no-go for life." For the discovery itself, Greaves's team used the James Clerk Maxwell Telescope in Hawaii, and the powerful Atacama Large Millimeter/submillimeter Array (ALMA) in Chile.

Phosphine is composed of one phosphorus atom bonded to three hydrogen atoms and forming a pyramid-shaped molecule. The key point is that while phosphine can be naturally created in the high-pressure environments prevailing, say, in gas giants such as Jupiter and Saturn, under the conditions typically existing on Earth or other terrestrial planets, its most common producers are anaerobic bacteria (which don't need oxygen to survive and grow), such as those existing in our intestines or in certain deep-ocean worms.

As often happens following the announcement of such a dramatically unexpected discovery, the tentative detection of phosphine on Venus immediately drew both great interest and considerable controversy, with other planetary scientists voicing a plethora of notes of caution. Several researchers questioned the phosphine detection itself, either on the grounds of data processing and analysis, or by suggesting that the observed signal should be attributed to sulfur dioxide rather than phosphine. Others raised doubts about the interpretation of the detection as representing a genuine biosignature. In 2022, researchers using the Stratospheric Observatory for Infrared Astronomy (SOFIA)—a far-infrared telescope mounted in a Boeing 747 aircraft—claimed they didn't see any sign of phosphine on Venus. Later, however, the apparent non-detection was attributed to calibration errors, and a low concentration of phosphine was in fact detected in the same dataset. While the authors of the original discovery have responded in detail to most criticisms, the debate on the reality of the detection continued. A new twist in the story appeared in 2023, when Greaves reported the discovery of phosphine deeper in the atmosphere than where it had been spotted before, down to the more temperate environment in the middle of Venus's clouds. It therefore seems that the presence (or not) of phosphine may be definitively resolved only through future attempts, perhaps by *DAVINCI*, to measure phosphine directly within the Venusian clouds.

The answer to the question of whether the detection of phosphine (assuming it's real) signals the existence of some form of life in the clouds of Venus is equally controversial. For example, origin-of-life researcher Gerald Joyce of the Salk Institute in California was quite explicitly skeptical: "This can hardly be taken as a biosignature," he stated, pointing out that the discoverers themselves noted in their paper that "the detection of phosphine is not robust evidence for life, only for anomalous and unexplained chemistry."

There are, however, other phenomena high in the Venusian atmosphere (some forty miles above the surface), which continue to intrigue researchers. For instance, there are some yet-unidentified substances, which very efficiently absorb ultraviolet radiation from the Sun. Those are reminiscent of the absorption provided on Earth's surface by photosynthetic pigments. There are also gases in the Venusian atmosphere, such as oxygen and methane, which appear to be in a state of thermochemical disequilibrium—a condition that on Earth is produced through life processes. These and a few other puzzling features have prompted a team of researchers from MIT, led by astrophysicist Sara Seager, to start a series of privately funded missions originally called the *Venus Life Finder* (*VLF*) and now called *Morning Star* missions. These will feature a series of direct atmospheric probes designed to assess the habitability of the Venusian clouds and search for signs of life and for life itself. One step in this quest will include a probe looking for signatures of life while descending through Venus's atmosphere. Estonia's Tartu Observatory is involved in the endeavor. It is building an instrument that will fly as one of the *Morning Star* Venus missions and is tentatively scheduled for launch by 2030. Upon arrival to Venus, the *TOPS* (*Tartu Observatory pH Sensor*) will dive into our sister planet's atmosphere to measure the acidity of single Venusian cloud droplets. Subsequent missions are planned to culminate in gathering samples and returning them to Earth.

The goal of the upcoming steps is clear: As with other contentious discoveries, the phosphine detection will first have to be confirmed by future observations, perhaps those performed via the *Morning Star* and *DAVINCI* missions. Following that, if the presence of phosphine is unambiguously confirmed, potential geochemical and photochemical sources of phosphine will have to be decisively ruled out, and a clearer identification of the anomalous ultraviolet absorbers will have to be achieved. Only then will we be able to even tentatively entertain an interpretation of these phenomena as being produced by an aerial microbial biosphere on Venus.

To sum up our thoughts about the possibility of finding life on the rocky planets in the solar system, here are a few humble opinions: We will be truly surprised if absolutely no signs of past life are discovered on Mars once the red planet is thoroughly explored. If that turns out to be the case, it would perhaps signal that even when the conditions are promising, the emergence of life may not be inevitable. We are somewhat more skeptical about current life on Venus, but surprises are definitely possible (especially in terms of unexpected chemistry), so exploratory missions should be encouraged. Journeying closer to the Sun, the blistering heat of Mercury, a planet that in itself is nothing but a pockmarked rock, does not make it hospitable to life.

Is there any other object in the solar system that could (in principle at least) support life? Surprisingly, as we shall see in the next chapter, there are at least three, and maybe even as many as six!

# CHAPTER 8

# Extraterrestrial Life on Solar System Moons?

*How many things have been denied one day,*
*only to become realities the next!*
—JULES VERNE, *FROM THE EARTH TO THE MOON*

The search for extraterrestrial life has largely been guided by a "follow the water" principle. The main idea behind this strategy is that even if nutrients and energy are available, without an adequate solvent, life's building blocks cannot be brought into contact to allow them to chemically react. In addition, water dissolves a wide variety of compounds that organisms consume, it can transport chemicals within cells, and it allows cells to dispose of waste. In general, water offers a complexity that opens up numerous reaction pathways that can lead to the emergence of life. Also, water has another fact going for it—on Earth, essentially wherever there is water, there is life. Following this line of reasoning, therefore, for a long time, rocky planets have only been considered "habitable" if they happened to fall within that circumstellar band of distances where the climate on their surface allowed for the stable presence of liquid water. In the solar system, only Earth (certainly), Mars (marginally), and Venus (optimistically) have been so lucky as to be considered potentially habitable. Our Moon and the Martian moons, Deimos and Phobos, are also formally in this habitable region, but the fact that they have sustained neither liquid water on their surfaces nor any atmosphere to speak of—the Moon, for example, does have an extremely dilute atmosphere composed primarily of neon, argon, and hydrogen, and it has water ice (near its poles and in shadowed craters) and small amounts of water molecules in the sunlit parts—has excluded them from the list of candidates to harbor life.

In recent years, however, we have discovered that the "habitable" universe may be, in principle, enormously expanded, to include, for instance, even some of the ice-covered moons orbiting the gas giant planets in the outer solar system. How is that possible? One reason is that it turns out that energy sources other than radiation from the host star—in particular, *tidal heating*—can result in a suitable temperature range to sustain large subsurface liquid oceans even at large distances from the central star. Another reason (albeit a speculative one) may be provided by the fact that on the surface of such distant moons, there can exist large lakes that are composed of liquids other than water. In other words, the concept of the "habitable zone" has ballooned far beyond its originally presumed real estate.

Here is, very briefly, an explanation as to how tidal heating works. When one celestial body is acted upon by the gravitational force of another body, such as when a moon feels the gravitational field of the planet around which it revolves, the moon is stretched a little along the line connecting the centers of the two bodies. This is a result of the fact that the point on the moon's surface that is closest to the planet feels a stronger gravitational pull than the moon's center, and the point that is farthest from the planet feels a weaker force than the moon's center. This stretching results in a "tidal bulge" being raised. If the moon's orbit is even slightly elliptical, the tidal force is strongest when the moon is closest to the planet and weakest when the moon is farthest. In other words, the orbiting moon experiences episodes of stretching and easing in each revolution, and those changes in the tidal deformation, as the bulge rises and falls, generate internal friction that heats the interior of the moon.

The story of the discovery that moons orbiting gas giants may conceal large liquid-water oceans underneath their surfaces provides for a fascinating demonstration of how advances in modern science can be achieved through step-by-step "detective" work.

It all started with spectroscopic observations of Jupiter's moons that were conducted in the 1960s and early 1970s. The great power of spectroscopy is that it can identify the composition of light-emitting material, or of matter through which the light passes. For instance, water ice gives a distinctive spectroscopic signature in the form of two absorption features in infrared light. Consequently, spectroscopy in the infrared readily revealed that Jupiter's moons Callisto and Europa were most likely covered with ice.

The next important step was taken by theorists. In a few seminal papers published in 1979 and 1980, planetary scientists Stanton Peale and Patrick Cassen from the University of California, Santa Barbara, and Ray Reynolds from the NASA Ames Research Center, proposed that tidal heating generated by the pull of Jupiter (and also by the so-called Galilean moons, Europa, Callisto, and Ganymede) could literally melt a significant part of the interior of Jupiter's closest moon, Io. The implication was clear: Io could exhibit volcanic activity and lava flows. Remarkably, a long series of subsequent observations, first by the *Voyager 1* and *Voyager 2* probes (in 1979, literally only a few days after the theoretical prediction was published!), and later by the *Galileo* spacecraft (in 1995 and in the 1999–2002 period), the *Cassini* spacecraft (in 2000), *New Horizons* (in 2007), and most recently by the *Juno* spacecraft (in 2023), have shown that Io is indeed the most geologically active object in the solar system. In particular, *Juno* passed within about 930 miles (1,500 kilometers) of Io in December 2023, and confirmed that Io hosts more than four hundred active volcanos. All of the discoveries so far mark only the beginning of a fascinating scientific adventure.

## Jupiter II

Io is certainly an extremely interesting object in its own right, but from the perspective of the search for life, it is believed to have the lowest amount of water (by atomic percentage) of any known object in the solar system. But Peale, Cassen, and Reynolds made another thrilling prediction, which may turn out to have dramatic consequences for the identification of potentially habitable worlds. They suggested that tidal heating could create a liquid-water ocean overlain by a crust of ice on Europa—the smallest of the four moons of Jupiter discovered in 1610 by Galileo Galilei (he referred to Europa as Jupiter II).

As intriguing as this prediction was, it also presented astronomers with a serious observational challenge: to confirm (or refute) the existence of such a subsurface liquid ocean buried under a many-miles-thick layer of ice. Faced with this difficult task, planetary scientists have creatively come up with a series of ingenious investigations.

First, they realized that there were already at hand a few suggestive clues. Images taken by *Voyager 2* showed that the surface of Europa is almost free of craters. In fact, it is the smoothest surface of any known solid body in the solar system. This implied the operation of some sort of natural Zamboni—just like ice skating rinks, Europa is being resurfaced by fresh ice. The resurfacing itself, in turn, suggested that ice movements, similar to plate tectonics, may be taking place. The most likely driver of such a dynamical geological activity could be a subsurface body of liquid. At the same time, Europa's surface was also found to be cracked—covered with dark streaks (dubbed *lineae* meaning "lines") crisscrossing it. These are reminiscent of oceanic ridges on Earth, again suggesting the equivalent of plate tectonics, with the cracks most probably caused by tidal flexing exerted by Jupiter's gravity. In addition, later spectroscopy by the *Galileo* spacecraft and by the Hubble Space Telescope revealed that the cracks appear to contain salts, which could have originated from an ocean underneath.

The second clue came from measuring the changes in the velocity of the *Galileo* spacecraft to the astonishing precision of a few tenths of an inch per second. Using this information John Anderson of the Jet Propulsion Laboratory and his team were able to map very accurately Europa's gravitational field, and thereby the distribution of mass density within the moon. They found that to fit the data, Europa had to be composed of three layers of different densities: an iron core measuring about 750 miles in diameter, surrounded by a rocky mantle made of silicates, on top of which there is a layer of water or ice, fifty to a hundred miles thick. The measurements were still not precise enough to be able to distinguish between liquid water and ice.

The final piece of evidence came from impressive magnetic measurements. The magnetometer (which measures changes in magnetic fields) on board the *Galileo* probe discovered that Europa acts like a weak bar magnet, where its magnetism is *induced* by Jupiter's magnetic field. Simply put, Jupiter has a strong magnetic field owing to its spinning hydrogen metallic core. Since Jupiter's magnetic axis does not precisely coincide with its rotation axis (the same is true for Earth), Europa experiences a periodically varying magnetic field (as if it is illuminated by a magnetic "lighthouse") as Jupiter spins around its axis. In electromagnetism, when an electrical conductor—material that allows electricity to flow through it—is placed in a varying magnetic field, the conductor itself becomes a magnet. In other words, the fact that the *Galileo* probe found that Europa has an induced magnetic field means that Europa contains a layer of electrically conductive material in its interior. From careful measurements of the induced field, Margaret Kivelson, Krishan Khurana, and their colleagues on the magnetometer team at the University of California, Los Angeles, were able to show that a salty, subsurface liquid ocean could indeed provide the implied conductivity.

The long and the short of this elaborate Herculean effort was that through a sequence of imaginative and careful examinations, researchers have been able to demonstrate that Europa's thick surface layer of ice most likely shields a salty ocean, which contains about twice the total amount of water as in Earth's surface oceans.

The depth of Europa's subterranean ocean is estimated to be on average about sixty miles, with a rocky seafloor underneath, and an outer crust of solid ice the thickness of which is uncertain, but guessed to be between a few and twenty miles. Interestingly, the Hubble Space Telescope may have provided one more piece of evidence for the presence of this ocean. An image of Europa taken by Hubble in 2012 revealed what appears to be a thin plume of water vapor rising to a height of about 120 miles. Researchers using Hubble also reported tentative detections of plumes in 2016 and 2017. Moreover, astronomers who had re-analyzed data from the *Galileo* probe concluded in 2018 that the spacecraft may have even flown through one such plume in its flyby of Europa in 1997. Finally, a research team using the Keck Observatory in Hawaii announced in November 2019 that they had directly detected water vapor above Europa's surface. These plumes, if unambiguously confirmed (by the fall of 2023 they had not been detected by the James Webb Space Telescope), may originate directly from the underlying ocean (through cracks in the ice), or from some pools or lakes of liquid water encased in the outer ice shell. The important point is that plumes could offer a way to analyze the ocean's composition without having to drill through miles of ice. A spacecraft could travel through the plume to sample and analyze it from orbit. In 2023 astronomers discovered another intriguing feature that indicates that the subsurface ocean and Europa's surface are linked. Using data from JWST, they have identified carbon dioxide in a specific region on the icy surface of Europa. Detailed analysis indicated that this carbon likely originated in the subsurface ocean and was not delivered by meteorites or other external sources. Moreover, it was deposited on a geologically recent timescale.

The answer to the question of whether Europa's salty ocean can really provide a potential habitat for life is no more than a "maybe," but that only means that we should definitely explore this possibility. The optimists base their hopes on data from several extreme environments that appear to support a diverse ecosystem in subglacial lakes on Earth. Chief among those is Lake Vostok in Antarctica.

Lake Vostok is buried under more than thirteen thousand feet of glacial ice, cut off from any contact with the atmosphere or light. It is estimated that it has been continuously covered with ice for about fifteen million years. In terms of size and volume, Lake Vostok is one of the largest freshwater lakes on Earth. Given its unique conditions, scientists from the United Kingdom, Russia, France, and the US have made several attempts to drill down through the ice almost to the lake's surface. In 2013, a team led by biologist Scott Rogers from Bowling Green State University performed DNA and RNA sequencing of frozen lake water attached to the bottom of the overriding glacier (dubbed *accretion ice*). That ice had been collected during drilling expeditions in the 1990s. Rogers and his colleagues identified thousands of gene sequences, mostly of bacteria and eukaryotes (organisms whose cells contain a nucleus), implying that the lake's water supports life. A few microbiologists voiced skepticism about the results, suggesting that the findings may have represented contamination from the drilling process itself, rather than genuine lake life. These doubts seem to have largely been laid to rest when microbial ecologist John Priscu of Montana State University obtained quite similar results at another subglacial lake in Antarctica—Lake Whillans—after drilling through 2,600 feet of ice. Scott Rogers himself further strengthened the veracity of his findings in 2020, when he and biologist Colby Gura published a new analysis of the biological diversity in Lake Vostok, basically confirming

Rogers's earlier results. They concluded their paper with: "Therefore, Lake Vostok may contain a functioning ecosystem that receives chemical and energy inputs from the overriding glacier and from possible hydrothermal sources."

There is another interesting aspect that we should consider when assessing Europa's potential ability to support life. As we have described in Chapters 3–5, experiments attempting to originate life in the laboratory have shown that the path from chemistry to biology (assuming such a complete path indeed exists) requires high concentrations of crucial molecules, to initiate and sustain the necessary chemical reactions. A large ocean environment is not ideal for obtaining such dense accumulations. This realization has led, in fact, to the suggestion (discussed in Chapter 6) that life on Earth started in relatively small ponds on land, rather than in the oceans. Does this mean, if true, that Europa could not harbor life? Not necessarily, although it may mean that life (if indeed present) did not originally emerge on Europa. As we have already noted, it is possible, for example, that rocks ejected (via asteroid or comet impacts) from Mars, where life could have (in principle) emerged, reached Europa. If such rocks happened to contain some forms of microbial life, then this life could (again, in principle) continue to be sustained and maybe even evolve in Europa's subsurface ocean. On the slightly more pessimistic side, a study published in March 2024 suggests that charged particles bombarding Europa's surface produce less oxygen (from breaking down frozen water) than previously thought.

The data that *Juno* has collected during its recent flybys are helping the researchers who are planning the *Europa Clipper*. The latter is a NASA mission scheduled to launch in October 2024 and to arrive at Jupiter in 2030. It is expected to make a series of flybys near Europa, a few perhaps as close as sixteen miles above the surface, while in orbit around Jupiter. The main goal of this mission is to confirm the existence of the subsurface ocean and to aid in choosing a site for a future lander mission.

Europa is not the only moon in the solar system that could, in theory at least, harbor life. Some may even argue that there are other moons that offer equal or better chances.

## The Little Giant That Could

Saturn's sixth largest moon, Enceladus, was named after one of the "Giants" in Greek mythology. These Giants, who fought the Greek gods for control of the cosmos, were not necessarily very tall, but they were known for their strength. For instance, the mythological giant Enceladus is depicted fighting Athena on a sixth-century-BCE dish from the Attic Peninsula in Greece (the dish is currently at the Louvre). The moon Enceladus was discovered by astronomer William Herschel in 1789, and was named by his son, astronomer John Herschel, in 1847.

You might not have expected a moon that is so far out in the cold and ranked only sixth in size even among Saturn's moons to attract much attention. After all, a moon like Enceladus that is just a little bit larger than three hundred miles in diameter is not anticipated to be able to retain any heat. But Enceladus presented astronomers with a puzzle right from the get-go. Images of the moon obtained in 1981 by the *Voyager 2* spacecraft, even though taken from a large distance, revealed that while Enceladus's southern part appeared to be rather smooth, almost free of craters, the northern part had considerably more impact craters. Since Saturn's moons are being continuously bombarded, this configuration indicated that somehow the area around the south pole of Enceladus had been resurfaced. Astronomers also knew that Enceladus was covered with ice, since, as Morgan Cable, a scientist at NASA's Jet Propulsion Laboratory, put it, Enceladus is "one of the whitest and brightest objects in the solar system."

Space research is intricate, it requires patience, and it is rather expensive. This is why following *Voyager 2*, researchers had to wait for more than twenty years just to take the next step in their exploration of

Enceladus. NASA's *Cassini* probe did finally fly by Enceladus between 2005 and 2017, and the spectacular views and the data that scientists were able to collect from a series of no fewer than twenty-three targeted flybys, were definitely worth waiting for. On its first approach, *Cassini* came to within just 725 miles of the moon, and its magnetometer detected a distortion in Saturn's magnetic field above Enceladus's south pole. This seemed to be caused by material being ejected from Enceladus, since the Cosmic Dust Analyzer on board the spacecraft detected many dust-size particles.

The results from subsequent flybys were nothing short of breathtaking. First, images revealed numerous plumes and geyser-like jets emanating from around Enceladus's south pole. It also became clear that the jets were feeding one of Saturn's rings, known as the E-ring. Second, as *Cassini* flew through the plumes and chemically analyzed their composition, researchers discovered that the plumes contained water vapor, carbon dioxide, carbon monoxide, methane, nitrogen, ammonia, and other carbon compounds, typically found in comets, but also in hydrothermal vents on Earth. A third stunning discovery was the finding that Enceladus's south pole was warmer than its equator, even though the pole receives less sunlight. Moreover, surface linear fractures observed around the south pole (which became known as "tiger stripes") had a temperature that was about seventy degrees Fahrenheit (thirty-nine degrees Celsius) warmer than the equator, definitely signaling some sort of geological activity in that region.

Based on the early results, the possibility that the jets and plumes were originating from a subsurface liquid-water ocean was already on the minds of all the scientists involved in the observations. There remained, however, even at that point, a nagging doubt, since there was still a possible alternative interpretation. The water vapor could be sublimating at the surface, directly from ice into gas, in a process similar to that observed in comets (albeit, in the case of Enceladus, without sunlight being the source of heat).

This was the stage at which the two teams behind the two instruments carrying out the chemical analysis on board *Cassini* were given the opportunity to shine. First, the Cosmic Dust Analyzer (CDA), which was operated by the University of Stuttgart in Germany, found sodium salts in Saturn's E-ring—a ring that was clearly created by the jets from Enceladus. In fact, a combination of data from *Cassini* and from the Herschel Space Observatory had shown that an ice-particle doughnut-shaped cloud around Saturn was also formed by the sprinkles from Enceladus. Given their source, the sodium salts in the E-ring meant that the jets themselves contained salts—a phenomenon that had never been detected in comets—making the surface sublimation scenario much less likely. In addition, the CDA detected specific silica particles, which on Earth are typically found at hydrothermal vents on the ocean floor. The silica crystals sampled from the Enceladus plume were between two and eight nanometers (billionths of a meter) in diameter, and grains of that size of pure silica are generally formed in a salty water ocean if they are exposed to temperatures of about two hundred degrees Fahrenheit—temperatures associated with hydrothermal vents. But at some hydrothermal vents on Earth molecular hydrogen is also produced at a very high rate. So there was a tentative prediction that the plumes should also contain molecular hydrogen. To everybody's delight, the *Ion and Neutral Mass Spectrometer* (*INMS*) instrument, run by the Southwest Research Institute in Texas, indeed showed that the plumes held much more molecular hydrogen than could have simply resulted from the breakdown of more complex molecules.

Even after this observational success, there were still two additional important questions that had to be answered. First, whether the subsurface ocean on Enceladus was global, or regional and confined only to the environs of the south pole. Second, whether this ocean was sufficiently old to have allowed for life not only to be sustained (once it had been delivered there), but perhaps also to evolve.

Fortunately, there was still a treasure trove of data from *Cassini* to be analyzed. By examining hundreds of photos from the entire duration of the flybys of the *Cassini* mission, scientists were able to map precisely the positions of features on the surface of Enceladus, and in 2015 they discovered that the moon wobbles in its motion. The importance of this discovery for mapping the internal structure of Enceladus can't be overemphasized. Enceladus is tidally locked to Saturn in the same way that our Moon is tidally locked to Earth. This means that it takes Enceladus the same time to spin around its axis as it takes it to complete an orbit around Saturn—resulting in the same side of Enceladus always facing Saturn (just as our Moon always shows us the same face). A small side-to-side wobble (known as *libration in longitude*) occurs when the moon is not perfectly spherical, so that the gravitational attraction is slightly out of whack. The degree of wobble can tell whether the moon spins as a solid body or whether its crust is in fact floating on top of a liquid layer. In the latter case the amplitude of the libration is larger. By comparing Enceladus's wobble with theoretical models, *Cassini* scientists concluded quite confidently that Enceladus contains a global ocean sandwiched between an icy surface layer and a rocky core.

Three new, exciting discoveries about Enceladus were published in 2023. First, using the James Webb Space Telescope, a team led by Geronimo Villanueva of NASA's Goddard Space Flight Center detected a water vapor plume from Enceladus spanning more than six thousand miles! Second, Frank Postberg from the Freie Universität Berlin and his colleagues analyzed individual ice grains emitted from Enceladus and found that phosphate is present in Enceladus's ocean at levels that are at least one hundred times higher than those in Earth's oceans. Recall that phosphorus, in the form of phosphates, is vital for all life on Earth. It forms part of the backbone of DNA and is also a component of some of the molecules of cell membranes. This study was the first to report direct evidence of phosphorus in an extraterrestrial ocean world. Third, scientists analyzing decade-old data taken by the *Cassini* mission from one of Enceladus's water plumes have taken the evidence for potential habitability one step further. They've found strong confirmation of the presence of hydrogen cyanide, a molecule that, as we have seen in Chapters 3–5, is key to life.

The detailed picture that emerged from all of the *Cassini* flybys has therefore transformed Enceladus from being a small, rather insignificant, distant moon, with no business harboring life, into one of the most promising astrobiology targets in the solar system. Enceladus is now thought to contain a subsurface salty ocean, underneath which there is an active seafloor with hydrothermal vents. The vents appear to involve a rich, molecular-hydrogen-producing chemistry—one of the types of environments that are colonized by life on Earth. As we have noted several times, we have reasons to believe that small ponds on land are more likely places for life to have originated, but hydrothermal vent environments certainly can sustain life. The evidence for the existence of hydrothermal vents, by the way, was further strengthened by the discovery of organic molecules, which are expected to be present at such vents.

One other finding from *Cassini* was particularly intriguing. An analysis of the composition of a plume through which *Cassini* had flown revealed a surprisingly large amount of methane. A team of researchers, led by Antonin Affholder of the École Normale Supérieure in Paris, showed in 2021 that the most likely abiotic chemical process—the reaction of minerals in rocks with $CO_2$ and hot water, known as *serpentinization*—would not have been capable of creating as much methane as had been observed. In contrast, the researchers have demonstrated that the action of methanogens, microbial life-forms that consume hydrogen and carbon dioxide and produce methane, could explain the detected amount of methane. But we shouldn't get too excited. The researchers were quick to acknowledge that there could be other solutions to the methane mystery, not involving life.

Those include, for example, an excess of primordial methane, left over from Enceladus's formation, or another still unknown abiotic process at work.

A subsurface ecosystem in Enceladus, if it indeed exists, would certainly not look like a garden-variety ecological community on Earth. Rather, it would have to be more akin to those habitats associated with extremophiles—microorganisms that are tolerant to environmental extremes. In particular, those organisms have to be anaerobic, and they should be able to thrive without relying on photosynthesis. Perhaps the closest Earth analogue is provided by the subglacial lakes in Antarctica and some environments in Iceland, with their abundant hydrothermal activity and jet-producing geysers.

The question of the age of Enceladus's ocean has not been settled. In a paper published in 2017, a team of scientists, led by aerospace engineer Luciano Iess from the Sapienza University of Rome, attempted to determine the age of Saturn's rings. The team's estimate was based on the amount of dust currently existing in the rings, and the average rate at which dust accumulates in the rings, assuming that it is coming in the form of micrometeoroids from the *Kuiper belt*—a region of icy objects beyond the orbit of Neptune. The age estimate obtained by Iess and his colleagues came as a surprise to many astronomers and planetary scientists—only about one hundred million years! In other words, if true, not only are Saturn's rings not as old as the solar system, but rather, by the time that they had formed, life on Earth had already evolved deep into the dinosaur era. This could have important implications for the prospects of Enceladus harboring life, since one of the scenarios for the formation of the rings suggests that both Enceladus, and Saturn's rings, were created simultaneously, as the result of Saturn (or one of its earlier moons) having been impacted by a relatively large object.

Not all researchers accept the conclusion of the analysis of Iess and his collaborators about the relatively young age of the rings. Specifically, a number of planetary scientists pointed out that the estimate of the rate at which the rings had been polluted by dust involved considerable uncertainties. Others argued that it would have been difficult to form the rings in such a short time. In fact, the mechanism for the formation of Saturn's rings has been debated by planetary scientists for many years. One of the latest of the formation models, presented in September 2022 by planetary scientist Jack Wisdom of MIT and colleagues, proposed that a former moon of Saturn (dubbed *Chrysalis*) was tidally torn apart some 160 million years ago to form the rings. An even more recent study by Sascha Kempf of the University of Colorado, Boulder, and collaborators determined the flux of micrometeoroid bombardment onto Saturn's rings and constrained the age of the rings to no more than about four hundred million years. Given the remaining uncertainties, however, to many astrobiologists, Enceladus endures as one the most attractive targets in the search for extraterrestrial life in the solar system.

## Maria and Lacus

In Latin, *maria* means seas, and *lacus* means lakes. There is only one object other than Earth in the solar system that has stable seas and lakes, rainy seasons, and even an Earth-like cycle of liquids on its surface—Saturn's largest moon, Titan. Indeed, images of Ligeia Mare, the second-largest body of liquid on Titan, taken by the NASA *Cassini* mission, could be mistakenly thought to represent a photo of bodies of water on Earth's surface taken from above. On July 8, 2009, *Cassini* even captured a *glint* (also known as *specular reflection*), the first flash of sunlight reflected off the Kraken Mare on Titan, confirming the presence of liquids in Titan's lake zone. Titan's atmosphere is thicker than that of Earth, and like Earth's atmosphere, it is nitrogen-rich. But don't be fooled by these seemingly Earth-like attributes into also expecting an Earth-type biosphere to exist on Titan's surface. The temperature characterizing the Titanic terrain is a forbidding minus 290 degrees

Fahrenheit (minus 179 degrees Celsius), which means that the lakes, seas, and rivers are not composed of liquid water, but rather primarily of liquid methane, ethane, and nitrogen. The rain that falls occasionally on Titan is composed of liquid methane drops as well. Titan's atmosphere also has its own peculiarities. It is composed of nitrogen (more than 95 percent), methane (less than 3 percent), hydrogen (a fraction of a percent), and just trace amounts of other hydrocarbons (organic chemicals made up of carbon and hydrogen). On top of that, it is filled with a thick, organic, orange-colored haze.

The presence of methane in the atmosphere was initially considered rather surprising, since the expectation was for it to be completely destroyed by ultraviolet light from the Sun in less than a hundred million years. The assumption is, therefore, that methane is somehow being replenished, either continuously or through episodic outbursts, emanating from underneath Titan's icy surface. This conjecture has gained considerable support from spectacular observations by the European Space Agency's *Huygens* probe, which parachuted in 2005 from the *Cassini* orbiting spacecraft all the way down to Titan's surface. During its descent, *Huygens* detected in Titan's atmosphere an isotope of the noble gas argon, which is produced by the radioactive decay of an isotope of potassium. Since the most likely location of this potassium isotope is embedded in rocks, the presence of argon in Titan's atmosphere strongly suggests that gases do escape from Titan's interior.

Surprisingly, Titan's strikingly unique surface and atmospheric characteristics are not the only interesting aspects of this moon. When it comes to considering it as a potential habitat for life, it has another promising feature—it also has a salty subsurface liquid-water ocean. The first hints for the existence of this ocean came from the discovery of a few, initially puzzling electromagnetic phenomena. The *Huygens* probe detected extremely low-frequency radio waves in Titan's atmosphere, as well as the presence of a nonzero electric field near Titan's surface. You may recall that in the case of Jupiter's moon Europa, the detection of an induced magnetic field implied the existence of conductive material within the moon. Similarly, detailed models of Titan, by planetary scientist Christian Béghin of the Université d'Orléans in France and colleagues, showed that the low-frequency radio-wave detection required the existence of a conductive layer beneath the surface—one whose properties were best matched by a salty ocean. Second, by tracing with great precision the orbit of *Cassini* between 2006 and 2011, researchers were able to characterize and map Titan's gravitational field with an accuracy that allowed them to make out its internal structure. In particular, by measuring the deformations that Titan experiences as a result of the changing tidal force exerted on it by Saturn, they showed that the moon was not reacting as a rigid body, but rather like one that has a fluid layer underneath its outer icy shell, again implying the presence of a liquid-water ocean. Neither the depth of this ocean nor the thickness of the surface ice layer are known with any certainty, but estimates by aerospace engineer and *Cassini* team member Luciano Iess and colleagues put both numbers at approximately fifty to sixty miles.

Titan offers (in principle at least) the fascinating possibility that two completely different types of life could exist on the same solar system object. One of those is life-as-we-know-it, inside Titan's subsurface ocean. The other, of life-as-we-don't-know-it, is in those lakes of liquid methane/ethane on Titan's surface. Titan therefore challenges our "follow the water" principle, which is entirely based on a biochemistry derived largely from the characteristics of water and water-based solutions.

We should note, though, that the reasons for our inclination to rely on water as a solvent for life don't stem only from an Earth-centric view—they have a solid biological-chemical-physical basis. First, the water molecule is polar, that is, it has small negative and positive electric charges on its oxygen and hydrogen ends, respectively. Polar solvents dissolve polar molecules, and many of the building blocks of life (as we know it) involve polar molecules. Liquid methane and ethane, on the other hand, are rather poor solvents. They are nonpolar, and while they can dissolve other nonpolar compounds such as *acetylene* and other hydrocarbons, they are not useful at all when it comes to the common components of carbon-based life. Second, water plays an important role in a variety of life processes, ranging from ensuring the structural stability of DNA and proteins, to protein folding. Third, the cells of all life as we know it are made primarily of water.

Still, an exciting possibility (in principle) is for Titan to host a completely different type of life, a true "second genesis." Astrobiologist Chris McKay from the NASA Ames Research Center thought that such life was realizable. He argued that the "free" organics found in Titan's atmosphere (for example, the photochemically produced hydrocarbon acetylene) could potentially serve as a source of chemical energy (primarily by reacting with hydrogen to produce methane and ethane). The question still remains, however, whether a liquid composed of methane and ethane can truly replace water as a solvent for life's molecules. As we have explained above, there is no doubt that liquid methane, with its extremely cold temperature, leaves a lot to be desired when compared to water. It is therefore not at all obvious that one can have much interesting chemistry take place in such an environment. Nevertheless, McKay's scenario did make one interesting, albeit speculative, prediction: if life were to exist in Titan's lakes, and if this life indeed involves the consumption of hydrogen, a depletion of hydrogen in the atmosphere near Titan's surface would be expected. Intriguingly, planetary scientist Darrell Strobel of Johns Hopkins University found in 2010 that *Cassini* data did indeed suggest that molecular hydrogen was flowing down through Titan's atmosphere, with the hydrogen essentially disappearing at the moon's surface. Another study, led by Roger Clark of the US Geological Survey in Denver, mapped hydrocarbons on Titan's surface and found a lack of acetylene—a result that is again consistent with McKay's speculative suggestion. Even so, most researchers, including McKay himself, caution that both the hydrogen and the acetylene results could have non-biological explanations. In particular, estimates of the expected biomass density (mass density in organisms) on Titan, admittedly involving many uncertain assumptions, suggest that the apparent decrease in acetylene is unlikely to have originated from processes involving living organisms. As Mark Allen, principal investigator with the NASA Astrobiology Institute's Titan team puts it: "Scientific conservatism suggests that a biological explanation should be the last choice after all non-biological explanations are addressed. . . . It is more likely that a chemical process, without biology, can explain these results—for example, reactions involving mineral catalysts." McKay correctly emphasized, however, that even the discovery of a non-biological catalyst that can be effective at Titan's frigid temperatures would in itself be quite startling.

Given Titan's truly unique features, we shouldn't be surprised to hear that this moon has inspired more than one speculation. For example, planetary scientists Ralph Lorenz of the Johns Hopkins Applied Physics Lab, Jonathan Lunine of Cornell University, and Catherine Neish from the University of Western Ontario proposed that the lakes of Titan could significantly improve their capabilities as solvents if they were to contain even tiny amounts of hydrogen cyanide. This possibility cannot be observationally excluded, since the relative abundance of hydrogen cyanide cannot be precisely determined on the basis of currently available data. The point is that since hydrogen cyanide is polar, the researchers suggested that the lakes could solvate (make a solute of ions with solvent) even polar molecules such as water. If this speculative idea is confirmed (first by precise laboratory experiments), the case for a potential habitat for life (of some form) on Titan's surface might be somewhat strengthened. We should note that hydrogen cyanide definitely exists on Titan. When the *Huygens* probe plummeted through Titan's atmosphere, it measured hints of hydrogen cyanide ice at around sixty miles above the surface. Also, in 2014, Remco de Kok, a planetary scientist at Leiden University in the Netherlands, and his team examined data gathered by the infrared spectrometer on board the *Cassini* mission and spotted several features in a cloud above Titan's south pole. They concluded that those features were produced by hydrogen cyanide ice. Curiously therefore, hydrogen cyanide, which was most likely crucial for the emergence of life on Earth, could, in principle, also be a facilitator of life on Titan.

Titan continues to provide a fertile ground for guesswork. For example, theoretical chemists Martin Rahm and Hilda Sandström from Chalmers University of Technology in Sweden suggested that cell membranes—one of the main characteristics of life as we know it—may be altogether unnecessary for a hypothetical astrobiology under the conditions on Titan. They proposed instead that life's molecules could simply rely on Titan's frozen environment to hold them together. Such molecules, they argued, could stay stuck to a rock, with nutrients floating their way providing them with a "free lunch."

While all of these imaginative ideas remain for now only at the borderline between science and science fiction (we shall discuss more possibilities of life as we don't know it in Chapter 10), NASA does plan to launch in 2026 the rotorcraft lander *Dragonfly* mission to Titan. This multi-rotor drone-like vehicle is expected to reach Titan in 2034, to fly to dozens of locations on Titan's surface, and to look for potential signs of prebiotic water-based or hydrocarbon-based chemical processes.

There are definitely other objects in the solar system that could in principle support simple life, although the case one can make for those is somewhat weaker perhaps than for the planets and moons that we have discussed so far. Beyond Europa, both of Jupiter's Galilean moons, Ganymede and Callisto, quite probably also contain subsurface salty oceans. Ganymede, whose diameter is about 41 percent that of Earth's, even possesses an intrinsic magnetic field. However, since the floors of these oceans seem to be composed of ice compressed under high pressure, rather than being rocky with hydrothermal vents, the chances of life having thrived there may have been smaller. Still, if life were to be delivered to these oceans via asteroid impacts, maybe it could have survived. The European Space Agency has developed a mission entitled the *Jupiter Icy Moons Explorer* (*JUICE*), which will study Ganymede, Callisto, and Europa. The plan is for the spacecraft to enter an orbit around Ganymede, in order to determine its internal structure. *JUICE* was successfully launched on April 14, 2023, and it is scheduled to reach Jupiter about eight years later. As part of the final preparations before launch, a commemorative plaque was mounted on the spacecraft, as a tribute to Galileo, who was the first to view and study Jupiter's four largest moons through his telescope.

There is some (although more limited) evidence for the potential existence of a subsurface ocean on Neptune's moon Triton, and even underneath the ice of the Sputnik Planitia, a nitrogen-ice-filled basin on the surface of the dwarf planet Pluto. Saturn's tiny moon Mimas also likely contains a young ocean underneath its icy crust. Similarly, images of the dwarf planet Ceres (taken in 2015 by NASA's *Dawn* mission) hint at geological activity. When combined with additional evidence for water vapor, and the putative detection of carbonates on Ceres's surface, the full complement of data suggests the possible presence of a subsurface ocean on this dwarf planet as well. Surprisingly, in 2024, a team of researchers found evidence for geothermal activity even within the icy dwarf planets Eris and Makemake, located in the Kuiper belt. Methane detected on their surfaces by JWST points perhaps to thermal processes producing methane in their rocky cores.

All of these additional objects are very interesting from a geological perspective in their own right, and they may provide important clues concerning the formation and evolution of the solar system, but we don't find them to be as promising candidates for harboring life as Europa, Enceladus, and Titan. The bottom line is simple: a few of the moons of Jupiter and Saturn should definitely be thoroughly explored. Even a failure to detect any life in the subsurface oceans of these objects (or the methane lakes on Titan) could provide important insights into the concept of habitability.

The discovery of thousands of extrasolar planets and planetary systems in the last three decades has emboldened astronomers to be more ambitious, and to expand their search for life beyond the solar system—into the Milky Way galaxy. The story of this search is the next step in our journey.

# CHAPTER 9

# Life Out There

## The Astronomical Quest

*If I cease searching, then, woe is me, I am lost. That is how I look at it—keep going, keep going come what may.*

—Vincent van Gogh, *The Letters of Vincent van Gogh*

If we were to find any form of life on Mars, Venus, or one of the moons in the solar system, this would undoubtedly be an extremely exciting discovery. However, unless that life truly exhibits a lineage completely independent from that of life on Earth—*a genuine second genesis*—there will always be the nagging suspicion that life in both places came from the same source, and that life was simply transported (a process dubbed *panspermia*) from one location to the other (for example, via some powerful asteroid impact). Consequently, there is no question that finding life on an extrasolar planet (exoplanet) in some distant planetary system would constitute a much more electrifying discovery, one that would have impactful implications far beyond science—it would literally change our perception of our place in the cosmos. In the near future, however, unless an extraterrestrial technological civilization physically visits Earth—something for which there is no convincing evidence yet that it has ever happened—a discovery of extrasolar life will only be feasible remotely, through the use of a variety of space-based and ground-based telescopes.

The first step has to be the detection of the exoplanets themselves. The major problem astronomers face in trying to observe exoplanets is that the stars around which these planets revolve are often millions and sometimes even billions of times brighter than the orbiting planets. For example, a star such as the Sun is a billion times brighter than any terrestrial planet orbiting it. Any light reflected off such a planet is completely drowned out by the powerful radiation coming from the planet's host star. To overcome this problem, astronomers have come up with ingenious observational techniques. Since we are specifically interested in the detection of life, we will not discuss in detail all the detection methods of exoplanets in general, but rather give a very brief outline of the primary methods that have led to the discovery of extrasolar planets, and then move on to the techniques that we hope will usher in the discovery of signs of life—*biosignatures*.

## Detecting Exoplanets

Most of the currently known exoplanets have been discovered by one of the following two methods: *transit photometry* or *radial velocity*. Transit photometry relies on the fact that if, as seen from Earth, a planet crosses in front of its host star (a phenomenon called a "transit"), then the observed flux of the star slightly decreases during the transit, typically by a percent or so. Radial velocity measurements are based on the physics of two gravitating objects revolving around their center of mass. Planets don't really orbit a stationary star. Rather, the star and its planet revolve around their common center of mass, being held together by gravitational attraction. This means that unless we see the orbit precisely pole-on, we will observe the star being nudged slightly (because its mass is typically much larger than that of the planet), but periodically, sometimes toward and sometimes away from Earth, due to the gravitational tug exerted on it by the planet. The radial (meaning, along the line of sight) velocity of this motion can be deduced from the periodic shift in the star's spectral lines, due to an effect known as the *Doppler effect*—light waves get compressed and therefore shifted slightly toward the blue, when the star is moving toward us, and stretched out and shifted toward the red, when it is moving away from us. Of course, it is easier to detect planets via radial velocities when the ratio is higher between the planet's mass and that of the star, since then the star is more affected by the gravitational pull of its planet, and the blue and red shifts are therefore larger. We should note that the radial velocities that astronomers measure are the projection of the actual velocity of the star, along the line of sight. Only when we see the orbit precisely edge-on, can we measure the actual value of the velocity. Consequently, when we apply one of the laws of planetary motion discovered by astronomer Johannes Kepler in the seventeenth century, measurements of the radial velocity generally yield only the *minimum* value that the mass of the planet can have. The radial velocity method can determine a value very close to the precise mass for *transiting* planets, since we can only see transits when we observe the orbit almost edge on.

Until the launch of the Kepler space telescope in 2009, the radial velocity method delivered the largest number of exoplanet discoveries. Since then, however, most of the exoplanet detections have been achieved by the transit method. By the end of 2023 more than four thousand exoplanets had been discovered using the transit method, compared to a little over a thousand through radial velocity measurements. Formally, we speak of a *transit* when a planet passes in front of the star (as seen from Earth), or, in general, when the smaller of the two objects passes in front of the other. When the larger object passes in front of the smaller one, we call it an *occultation*. Obviously, not all planets exhibit transits, since transits occur only when the plane of the orbit aligns perfectly from the observer's vantage point. One can easily show that since the inclination of planetary orbits to our line of sight should be randomly distributed, the probability of observing a transit in a star-planet system is approximately given by the ratio of the radius of the star to the radius of the planetary orbit for circular orbits (or the semimajor axis of the orbit for elliptical orbits). This means, as could be expected, that close-in planets have a higher probability of producing observable transits. On average, for planets on close-in orbits, the probability of a transit is about 10 percent, and this probability decreases the larger the orbit. For instance, for a planet orbiting a Sun-like star in a circular orbit with a diameter equal to that of Earth's orbit around the Sun, the probability of a transit is less than half a percent. Consequently, surveys searching for transiting exoplanets had to scan hundreds of thousands of stars. This gigantic effort has led to thousands of detections. During its 9.6 years in orbit, the *Kepler* mission discovered more than 2,660 exoplanets, by observing more than half a million stars.

The amount of dimming that is observed during a transit is equal approximately to the fractional area of the stellar face that is being blocked by the planet. Therefore, astronomers can determine the radius of the exoplanet from the observed reduction in flux, since typically they know the radius of the star from its observed luminosity and surface temperature. As we have noted above, for transiting exoplanets, the geometry implies that the orbits are being observed nearly edge-on. Radial velocity measurements therefore give the planet's mass, and the mass and the radius together determine the planet's mean density, since the density is equal to the mass divided by the volume. Knowledge of the density is key in the search for life, since it allows for the identification of Earth-like planets—the mean density of rocky planets such as Earth or Mars is higher than that of gas giants like Jupiter or Saturn. Interestingly, observations of the *TRAPPIST-1* system, an amazing collection of seven "Earth-sized" planets orbiting a red dwarf star some forty light-years away, have shown that all seven planets are rocky and have very similar densities.

A drawback of the transit method is the danger of a relatively high rate of false detections. There are three main sources for potential false discoveries. First, there are eclipsing binary stellar systems in which the eclipsing star only just grazes the limb (outer edge) of the other star (as seen from Earth), thereby producing a very small dip, mimicking a planetary transit. Second, there are eclipsing binary systems in which the coincidental presence of a third star along the line of sight gives the impression of a shallower (than the actual) eclipse depth, again resembling a planetary transit. Third, white dwarf stars—the dense remnant cores of Sun-like stars—are about the same size as planets (as can be brown dwarfs, which are objects that failed to ignite hydrogen-burning nuclear reactions at their centers), and therefore they give rise to a similar dimming to that produced by giant exoplanets. In all of these cases, astronomers need to use additional observational data to weed out the false detections.

Other methods used to detect exoplanets include, for example, *gravitational microlensing*, where the gravitational field of a star and its planet magnify the light of a background star to produce a characteristic light curve. By the end of 2023, more than two hundred exoplanets had been discovered this way. Another method is *astrometry*, which consists of precisely observing minute changes in a star's position on the sky as it wobbles around the center of mass due to the presence of a planet, but only a couple of planets have been discovered so far using this method. A method that will undoubtedly gain more prominence with future telescopes is *direct imaging*, where planets are detected through their own thermal emission. With the currently existing technology, about seventy exoplanets had been discovered through direct imaging by the end of 2023. Another interesting method is *pulsar timing*—anomalies in the timing of the observed radio pulses from spinning neutron stars around which planets revolve. This method has been used to discover more than half a dozen such unusual planets.

As we have emphasized, however, here we are primarily interested in methods that can lead to the detection of signs of life, and *imaging* will most likely become a powerful tool with the next generation of large telescopes. For example, space telescopes equipped with coronagraphs will be able to block the light from the central star just like you might use your hand to shade your eyes from bright sunlight, making it possible to image planets. There are also proposals to fly a *starshade* to such a location in space (between the observed star and the space telescope) that it will prevent the blinding starlight from entering the telescope.

Discovering the exoplanets is only the first step. The main goal is to identify which exoplanets are habitable, or better yet, if any of them truly harbors life.

## Habitable Exoplanets

Most astronomers would agree that the simplest defining criterion for a planet to be considered "habitable" is its ability to sustain stable, long-lasting liquid water on its rocky surface. In other words, if we could *directly* detect the presence of a body of water on the surface of an exoplanet, that planet would automatically be considered habitable. Such a detection could perhaps become feasible with the next generation of space telescopes.

The idea behind observations that have been proposed with this goal in mind relies on the fact that when viewed at indirect angles, oceans (and lakes) reflect light differently than land, producing flashes of light—a phenomenon known as *glint*. You may recall that such a glint has already been observed from a lake on Saturn's moon Titan (although that lake contains liquid methane and not water). In addition, as the exoplanet rotates around its axis and at the same time revolves around its host star, we would observe different parts of the planet's surface, and those would be illuminated at different angles by the star (a bit like the phases of the Moon). Since oceans and lakes are more reflective than land, by using simulated planets and carefully analyzing the light that would be observed from them, researchers have demonstrated that they can construct maps of the surface reflectivity (a property known as the *albedo*), and thereby discover whether oceans are present. Comparing their simulated results with actual observations of Earth from a distance by NASA's *EPOXI* mission, scientists from the University of Washington concluded that future-planned space telescopes, with diameters upward of six meters (about twenty feet), could measure glint effects for between one and ten exoplanets orbiting the nearest Sun-like stars or smaller in the habitable zone. Researchers have also developed data analysis tools that could potentially detect both ocean-land heterogeneity and varying cloud cover, by mapping planet surfaces using direct imaging observations (again with future telescopes).

In the absence (so far) of direct observations of oceans, astronomers rely on other methods to determine which exoplanets can be considered habitable. Those methods ideally include observational determinations of the atmospheric temperature and pressure coupled with the detection of atmospheric water vapor. More crudely, a rocky (terrestrial) exoplanet may be considered habitable if it is in the estimated habitable zone of its host star. We should note that astrobiologists have even attempted to identify "superhabitable" planets—meaning star-exoplanet systems that could in principle allow for planets to be even more suitable for life than our Earth. In the description in this chapter, we are not considering the possibility of life on moons orbiting extrasolar giant planets, since detailed observations of such moons outside the solar system (*exomoons*) are beyond the capabilities of existing telescopes. In fact, to date there isn't even a single unambiguously confirmed exomoon, even though a few candidates have been detected (for instance, in November 2022, an exomoon candidate was reported around the planet Kepler-1513b). We also exclude from consideration the potential for life in subsurface oceans of ice-covered exoplanets, because the remote detection of such life in the near future is unlikely.

There are two other factors that determine whether we can expect to detect life on an exoplanet. One is the theoretically anticipated lifetime of the central star. If life on Earth can serve as a crude guide, then the host star (which is the primary energy source for life) needs to live for at least a couple of billions of years for *detectable* biosignatures to have time to evolve on an orbiting planet. For example, a high concentration of oxygen in the atmosphere of an exoplanet is considered a good (although not conclusive) biosignature. However, the detection of biogenic oxygen requires two things: first, oxygenic photosynthesis, and second, time for the produced oxygen to oxidize all the surface iron before it can start increasing in concentration. Also, the burial of reduced (organic) carbon may be required for oxygen to build up to high concentrations. Massive stars burn through their nuclear fuel furiously. Consequently, the more massive the star, the shorter its lifetime. For instance, while the lifetime of our Sun in its stable, hydrogen-burning phase is about ten billion years, the lifetime of a star with ten times the Sun's mass is only about twenty million years. We can therefore expect to find active biospheres only around stars less massive than about 1.5 solar masses.

There is a second element that can determine habitability, even for terrestrial planets in the habitable zone, and it's related to the question of whether or not there is a *lower limit* to the mass a star can have and still be able to host a planet that is harboring life. This question is particularly important since M-dwarf stars, with masses between 0.08 and 0.5 solar masses, represent about 70 percent of all stars in the Milky Way (even though these stars are so faint that from Earth none of them are visible to the naked eye). These low-mass stars are also the ones that live longest. A star with half the mass of the Sun is expected to live for almost sixty billion years, and a star with one-tenth the mass of the Sun for trillions of years. Hence, these stars can offer, in principle at least, the longest timespan for the emergence and evolution of life. Moreover, astronomers speculate that perhaps even as many as 80 percent of M-dwarfs may have planets in their habitable zones.

Yet, in spite of their seemingly positive attributes vis-à-vis the possibility of life, several concerns have been raised over the suitability of M-dwarfs as hosts to life-bearing planets. Here are just a few of those. First, M-dwarfs often experience intense flaring activity—eruptions of radiation that also trigger mass ejections. The flares increase in both frequency of occurrence and in amplitude the smaller the mass of the star. Moreover, even M-dwarf stars that are relatively less active exhibit flares and episodes of significant ultraviolet and X-ray emission during the first billion years or so of their evolution. Planets exposed to these harsh events could lose their atmospheres, their oceans, or both, and be totally sterilized by the time the star has settled into its long hydrogen-burning phase (the so-called *main sequence*). Second, since the habitable zone around M-dwarfs is snuggled much closer in to the central star (due to the low luminosity of these stars), planets are likely to be tidally locked by the gravitational interaction, presenting only one face to their parent star (just like our Moon is tidally locked to Earth and Enceladus is tidally locked to Saturn). This could generate a huge temperature difference between the permanent-day and perpetual-night sides of the exoplanet, with gases on the night side frozen solid, while the day side is being starbaked dry. As a result, scientists used to be quite skeptical about the ability of tidally locked planets to host a biosphere. If life existed there at all, they argued, it would have had to be confined to that eternal, narrow twilight zone around the line separating the two sides.

Opinions on the habitability of planets orbiting M-dwarfs started to change somewhat in recent years. First, a few researchers suggested that magnetic fields could shield exoplanets from the deleterious effects of flares, and prevent a significant erosion of the atmosphere and oceans by mitigating against the effects of stellar winds. Second, theoretical simulations identified atmospheric mechanisms that could, in the case of sufficiently dense atmospheres, circulate and distribute heat from the day side to the night side. For example, strong winds could be expected, with hot gas gusting at high altitudes toward the cold side, and cold winds blowing at low altitudes toward the side facing the star. In addition, modeling by a group of researchers from the University of Exeter in the UK showed that airborne mineral dust could cool the day side and warm the night side of a tidally locked planet, thereby broadening the habitable area. These researchers also identified a plausible feedback mechanism that could increase the amount of dust in the atmosphere and delay the loss of water from the day side (on planets at the inner edge of the habitable zone), thus keeping these planets habitable for longer periods of time. As a result of these and similar theoretical ideas, and perhaps also because of the "it's easier to search under the lamppost" effect—the fact that many of the discovered habitable-zone exoplanets orbit M-dwarfs, and those are the first to be studied—this class of objects has become a primary target in the search for extrasolar life. In particular, the hope has been that the James Webb Space Telescope (JWST) might answer at least the important question of whether exoplanets orbiting M-dwarfs have given rise to a stable atmosphere at all. Since potential biosignature signals from rocky planets around M-dwarfs are expected to be very faint, even JWST may not give us much more information. As University of Chicago astronomer Jacob Bean puts it: "They [observations with JWST] may be enough to tell us that an atmosphere is present, but the likelihood of them telling us anything about biosignatures on these rocky exoplanets—I really don't think that will happen."

Surprises are always still possible, and a few astronomers are somewhat more optimistic. Astrophysicist Sara Seager of MIT, for example, hopes that JWST will be able to determine whether some of the habitable-zone exoplanets around M-dwarfs have water vapor in their atmospheres (which, in turn, may imply that there is liquid water on their surface). Irrespective, however, of what precisely JWST will discover, there is no doubt that this powerful telescope will not only shine a brighter light on the enormous diversity of exoplanets, but it will also give us a first glimpse into the atmospheres of potentially habitable planets (or at least tell us whether such atmospheres are present).

## Proof of Life

Suppose that we discover an Earth-like planet in the habitable zone of a Sun-like star or an M-dwarf. How will we be able to tell whether it harbors life or not? In addition to a long list of technical difficulties involved in the attempts to detect life (which we shall shortly describe), there are also obstacles related to the evolution of life itself. For instance, suffice it to note that someone who had observed Earth three billion years ago, with telescopes similar to those that we now possess, would most likely not have identified any signs of life, even though life on Earth had emerged more than three and a half billion years ago. The point is that it is impossible to detect any unambiguous biosignatures before life has had sufficient time to modify the planetary atmospheric and surface environments to such an extent that those changes can be clearly discerned through telescopic observations. Such a major impact of life on its habitat is part of what has been dubbed "niche construction." To be clear, when we talk about biosignatures on exoplanets we mean any characteristics, substances, or features that can be used as evidence for past or present life. By their very nature, any remote-sensing signatures short of the unequivocal detection of an intelligent, technological species (technosignatures), will most likely be only *suggestive* of the presence of life.

We have already mentioned one of the best examples of ecological niche construction on Earth—the enrichment of Earth's atmosphere with oxygen, through the process of oxygenic photosynthesis by tiny organisms known as cyanobacteria (or blue-green algae). Remarkably, until about 2.4 billion years ago, Earth's atmosphere contained no oxygen to speak of. Around that time, the concentration of oxygen increased by several orders of magnitude during a relatively short period of time—an episode known as the "Great Oxidation Event." A subsequent major increase, one that brought the oxygen concentration to its modern levels, only happened between 750 and 460 million years ago. In fact, the emergence and diversification of large, multicellular life-forms on Earth might have had to await this second rise in oxygen, in both Earth's oceans and its atmosphere.

Another, rather more disturbing example of the effects that life can have on the planetary environment is what most scientists believe to be the human-activity-induced climate change that we are currently experiencing on Earth, and the associated decline in biodiversity, and increase in ocean acidification. It appears that for the first time in the 4.5 billion years of Earth's existence, humans may be able to determine the future of Earth's biosphere. To attach some numbers to the effects on acidification and biodiversity, the acidity in the oceans has increased by about 30 percent in the mere 250-plus years that have passed since the Industrial Revolution began, because of the increase in atmospheric carbon dioxide. Biodiversity is the variety of life on Earth in all its forms, comprising the number of species, their genetic variation, and their interaction within complex ecosystems. In a United Nations report published in 2019, scientists warned that up to a million species (out of an estimated 8.7 million) are threatened with extinction.

One of the many major challenges in detecting life on extrasolar planets is the weeding out of "false positives"—apparent biosignatures that are in fact generated by processes not involving life (*abiotic*). We have already encountered this type of difficulty in the claimed discoveries of life on Mars and on Venus, but the problem is obviously much more acute when one is dealing with relatively limited telescopic data, obtained from observations of distant exoplanets. Consequently, rather than reviewing all the potential biosignatures and the entire set of observational techniques that have been considered, we shall concentrate on those few that in our view offer the best chances to lead to a successful discovery of extrasolar life (assuming it exists) in the relatively near future.

The idea that one could discover life on other planets via the detection of gaseous biosignatures is not new. As early as 1965, James Lovelock, the originator of the "Gaia hypothesis," which proposed that living organisms form a synergetic system with their inorganic environment, published an article entitled "A Physical Basis for Life Detection Experiments." In this paper, Lovelock presciently identified gases that are in chemical disequilibrium with the rest of the atmosphere as one of the signs of life. He wrote: "Search for the presence of compounds in the planet's atmosphere which are incompatible on a long-term basis. For example, oxygen and hydrocarbons co-exist in the Earth's atmosphere." Molecular biologist and artificial intelligence researcher Joshua Lederberg also published in the same year an article about possible ways to detect biosignatures, with a special emphasis on Mars.

The concept behind searching specifically for gaseous biosignatures is simple. Metabolism is one of the characteristics of life, and it produces by-products that can, over time, accumulate in the atmosphere. If we can find a way to detect such gases in the atmospheres of exoplanets, then those planets could be, in principle at least, harboring life.

Theoretically, identifying the types of gases that might indicate the presence of life is straightforward: either gases that cannot be produced through purely abiotic chemical reactions, or gases that cannot remain in thermochemical equilibrium in an atmosphere for long periods of time. In the latter case, if such gases are identified in the atmosphere of an exoplanet, it means that they are continually being copiously produced, and the most likely source currently known to create them is life itself.

To determine which specific gases to look for, planetary scientists, atmospheric scientists, and astrophysicists have adopted two distinct, but complementary, strategies. First, given that Earth itself harbors life, they have treated Earth as if it were an exoplanet, and examined in detail spectra of Earth's atmosphere. This was achieved on one hand by spacecraft that turned back to observe Earth from afar (for example, the *EPOXI* mission observed Earth for several months from a distance of thirty-one million miles), and on the other, by analyzing *Earthshine*—the glow that lights up the unlit part of the Moon due to the Sun's light reflecting off the illuminated Earth's surface and onto the Moon.

As a second plan of action, researchers have run thousands of computer simulations of model exoplanetary atmospheres, in which they examined and followed in detail the evolution of the atmospheric chemistry, for a wide range of compositions and host star types. Naturally, the scientists attempted to ensure that the reaction network included both all the main processes that could generate traceable molecules in the atmosphere, and those that could destroy and remove such molecules.

From all of these intensive research endeavors, researchers came up with an "A list" of the best candidate biosignature gases. This list is rather short—it includes *molecular oxygen*, *ozone*, and *nitrous oxide*, with *water* and *methane* playing important supporting roles.

The most extensively studied among biosignature gases is molecular oxygen (composed of two atoms of oxygen) and, to a somewhat lesser extent, the closely associated ozone (composed of three atoms of oxygen). The fact that oxygen is so high on the list of signs of life is not surprising. First, on Earth, oxygen has been produced almost exclusively as a metabolic by-product by photosynthetic bacteria and plants. Second, oxygen has become a dominant component of Earth's atmosphere (21 percent by number of molecules). Indeed, the concentration of oxygen in Earth's atmosphere is about a hundred million times higher than that one would have anticipated from pure (abiotic) equilibrium chemistry. Third, and very important for the actual search for life, oxygen exhibits spectral signatures that are in principle detectable remotely. For example, molecular oxygen produces a relatively strong band in reflected light that is in the visual-near-infrared part of the spectrum, at a wavelength of 0.76 microns (one micron is a millionth of a meter, about 0.00004 inches). This spectral band has the additional advantage that it does not overlap with features from other common gases. Finally, an additional reason that makes oxygen a very promising biosignature is the fact that simulations have shown that on Earth-like planets revolving specifically around Sun-like stars, we do not expect oxygen to build up appreciably from processes that do not involve life (such as geological, or photochemical—where light dissociates water or carbon dioxide molecules, thus releasing oxygen). In other words, if we were to detect on such an exoplanet a high concentration of oxygen (say, at the 20 percent level), this would constitute a very encouraging sign of life. We should caution though that several studies that have simulated the atmospheric chemistry of potential alien worlds have shown that abiotic oxygen could accumulate on planets revolving around other types of stars (in particular, M-dwarf stars). For instance, on exoplanets experiencing very high evaporation rates of oceans because of a runaway greenhouse process, the water vapor could enter the exoplanet's stratosphere, where the vapor could be dissociated by radiation from the central star (with the hydrogen being lost to space), resulting in a relatively high oxygen concentration. Another abiotic mechanism that has been identified as capable of generating Earth-like accumulations of molecular oxygen is the splitting of liquid water on the planet's surface itself, through the action of titanium oxide as a catalyst. Detailed atmospheric studies, therefore, correctly highlight the danger of "false positives"—meaning that the detection of molecular oxygen *alone* (or ozone, see below) cannot be taken as definitive evidence for life. At the same time, the abiotic production of molecular oxygen or ozone may leave behind other clues, which could identify it as a non-biological process. For example, if the oxygen is indeed released from the splitting of water vapor molecules, this could result in a suspicious paucity of water vapor.

As we have noted above, thermochemical disequilibrium is another suggestive sign of life. A study aimed at investigating which particular disequilibrium in Earth's atmosphere is the most extreme—in terms of the power (rate of energy input) required to generate such an out-of-equilibrium state—identified the simultaneous coexistence of large concentrations of molecular oxygen, molecular nitrogen, and liquid water as the most unbalanced indicator. Without the action of living organisms, the oxygen, nitrogen, and water would have participated in a series of reactions with the ensuing equilibrium result of producing nitrates (salts containing a group composed of an atom of nitrogen and three atoms of oxygen) in the oceans. Consequently, an exoplanet in which these three components are identified (liquid water and high concentrations of oxygen and nitrogen) would be a very strong contender for harboring life. In principle, such detections would be feasible with the next generation of telescopes. Oceans could be revealed through the glint effect, molecular oxygen via its absorption features in near-infrared light, and molecular nitrogen by an absorption feature at a wavelength of 4.1 microns, resulting from the collisions among nitrogen molecules. We should emphasize, though, that the fact that this combination of signatures is potentially detectable does not automatically imply that such a detection is easy or imminent. Rather, it simply represents a worthwhile challenge for the relatively near future.

Ozone is another promising biosignature. The ozone in Earth's stratosphere is produced by photochemical reactions in which ultraviolet radiation from the Sun splits molecular oxygen, and the resulting atomic oxygen combines with molecular oxygen to form the triple-atom molecule of ozone (typically with the aid of a nitrogen molecule that absorbs energy). What makes ozone an attractive biosignature gas in its own right (in addition to the fact that molecular oxygen is needed for the formation of ozone) is the fact that its spectral signature as it absorbs energy occurs in the ultraviolet and mid-infrared parts of the spectrum, hence not overlapping with the prominent bands of molecular oxygen.

Another good candidate biosignature gas (at least for life as we know it) is nitrous oxide, which is composed of two atoms of nitrogen and one of oxygen. On the modern Earth, this gas is typically produced by life through microbial reactions. Very few non-biological sources produce nitrous oxide (although lightning does), but these sources usually create only extremely small concentrations. While on exoplanets revolving around very magnetically active stars, ultraviolet light could create more nitrous oxide, it would be possible (in principle at least) to distinguish this production from the biological one, since an abundant presence of other associated and detectable compounds (such as nitrogen oxide) would be expected.

Methane is also often considered and discussed in the context of the search for life (as we have already seen in the case of the observations of, and experiments on, Mars). Methane is the most stable (from a thermodynamic perspective) carbon-containing molecule in reducing atmospheres (for example, those that are dominated by hydrogen). We should recall though that in addition to it being produced by single-celled organisms (methanogens), methane can also be created by a variety of non-biological processes, as in the case of Saturn's moon Titan, where it is abundant. But methane is still very useful, because if one finds methane and oxygen or ozone together (or methane together with other oxidizing gases), this is fairly strong evidence for life. The reason that the simultaneous presence of methane and oxygen molecules is a quite reliable signature of biological activity is that methane cannot last long in an atmosphere containing oxygen-bearing molecules—no more than the contemporaneous existence at the same location of a large body of students and sizable quantities of Oreo cookies is likely to be sustained. Rather, methane and molecular oxygen rapidly oxidize to carbon dioxide and water. This means that if both are observed concurrently, the methane has to be continually freshly replenished, and the easiest way to restore the methane in the presence of oxygen is through life processes. The opposite is also true. In order to keep the oxygen around in an atmosphere that is replete with methane, one has to constantly bring back the oxygen, which is best achieved via reactions involving life. The simultaneous presence of carbon dioxide (which typically implies a neutral to oxidizing atmosphere) and methane would also indicate the possibility of life, since the methane would have to be produced either via biological processes, or through reactions between rocks and water—indirectly suggesting at least the existence of liquid water on the planetary surface. As we have discussed in earlier chapters, large asteroid impacts can generate a transiently reducing Miller-Urey type of atmosphere. This would be transient, but could last up to millions of years, depending on the size of the impactor. Detecting such atmospheres would be extremely exciting from the point of view of prebiotic chemistry, even though such reducing atmospheres might provide confusing signals with regard to life detection.

Unlike in the case of Mars, however, where simultaneously one rover was traveling on the Martian surface and another probe was orbiting the planet, detecting methane on exoplanets, even by itself, is not easy. Its strongest spectral band in infrared light is centered between seven and eight microns, where it overlaps with spectral features of both nitrous oxide and water. Consequently, a convincing detection of methane on exoplanets will require telescopes with a high spectral resolution. On top of that difficulty, the history of Earth's atmosphere suggests that the simultaneous detection of oxygen and methane may be somewhat of a chimera. On present-day Earth the concentration of methane is so low (about 1.6 parts per million) that its spectral signature would be undetectable. On the other hand, on the very early Earth (for example, during the *Archean* period, between 2.5 billion and 4 billion years ago), when the methane level might have been higher, oxygen was totally absent. A better indicator for potential life-forms during the Archean would probably have been the simultaneous detection of carbon dioxide and methane (perhaps accompanied by liquid water and nitrogen).

Finally, we should mention the interesting possibility that photochemistry involving high concentrations of methane can form a plausibly detectable *organic haze*, similar to the orange haze we observe on Saturn's moon Titan. Such a haze absorbs ultraviolet light, producing a distinctive feature in the observed reflected light. While the presence of an organic haze in itself certainly does not imply the existence of life, it can nevertheless single out potentially habitable exoplanets which are worthy of further examination. Indeed, hazes can also form following large impacts, with the resulting transient non-equilibrium atmospheric states.

---

Because of the risk of false positives, no single gaseous biosignature can be taken as definitive evidence for life. Astronomers have therefore considered a few other gases, beyond those most promising candidates, which could at least be suggestive of life. Examples in this category include gases containing sulfur. Such gases are of secondary importance as biosignatures since whereas it is true that metabolism directly produces sulfur-containing gases such as hydrogen sulfide and sulfur dioxide, volcanic and hydrothermal processes create these particular gases in much greater abundance. More interesting perhaps in the context of biosignatures are certain minor sulfur-containing gases. Studies have shown that ultraviolet radiation from certain types of stars could catalyze the copious production of ethane from more complex organosulfur compounds (such as *dimethyl sulfide* [DMS] and *dimethyl disulfide*). An anomalously high ethane signature could, therefore, suggest a sulfur-rich biosphere. In addition, since ethane has quite strong spectral features in mid-infrared light, its detection is definitely feasible. As we shall see later in this chapter, a tantalizing (but so far very tentative) potential detection of dimethyl sulfide on an exoplanet was reported in September 2023. On Earth, the bulk of DMS in the atmosphere is emitted from phytoplankton in marine environments (but more work needs to be done on abiotic ways of making it).

Researchers have also examined the chlorine-bearing gas *methyl chloride*. This gas is typically produced by a variety of plants and algae and by decaying organic matter. However, it can also be generated by abiotic processes, such as volcanism. The main feature that makes methyl chloride perhaps less attractive is the fact that in most cases it cannot survive for very long in an exoplanet's atmosphere—it reacts strongly with hydroxide (the component of water consisting of a hydrogen atom bonded to an oxygen atom). The only places where methyl chloride can perhaps be expected to be found are the atmospheres of planets around small red dwarf stars (M-dwarfs) that also happen to lack flaring activity (and consequently no photodissociation of water occurs).

Life may also alter a planetary environment in ways other than producing gaseous biosignatures in the atmosphere. For example, life could change the reflection and absorption properties of the planet's surface. It could also produce seasonal or other time-dependent variations in the observed spectrum of a planet. In the following section, we shall briefly discuss a few of these potential *surface* signatures of life.

## Life on the Edge

The surface biosignature that has attracted the most attention is caused by vegetation, and is known as the *Vegetation Red Edge* (*VRE*). The VRE represents the fact that there is a very steep change in the reflectance of vegetation in the near-infrared part of the spectrum. The chlorophyll in plants absorbs most of the visible light (for example, blue light and the red light in the wavelength range of 0.66–0.7 microns), but plants scatter and reflect near-infrared light (in the range of 0.75–1.1 microns). Consequently, the reflectance of vegetation can increase in an almost step-like fashion, from being around 5 percent at a wavelength of 0.68 microns (visible red light) to about 50 percent at 0.76 microns (near-infrared light). This is the reason, by the way, for why foliage appears very bright in infrared photography, and why the VRE is often used by Earth-observing satellites to examine the conditions of forests and vegetation.

Estimates show that the detection of the VRE on Earth-like exoplanets is still going to be extremely challenging, because even for telescopes with a relatively high resolving power it would require the exoplanet not only to have a significant fraction of its surface covered with plants, but also to be cloud-free over the vegetated areas for a non-negligible part of the day (on the average). This last condition may not be trivial to satisfy, since at least in the case of Earth, data from geostationary satellites suggest that there is a strong increase in cloud cover over large forested regions.

While the VRE is definitely a bona fide biosignature (since no abiotic sources for false positives are known), the fact that it is based specifically on the properties of vegetation on Earth makes its applicability to exoplanets somewhat uncertain. It is possible, for instance, that substances other than chlorophyll would produce "edges" or other types of signatures at different wavelengths (other edge features may also be more prone to false positives).

Researchers have also examined *chirality* as a potential surface biosignature—the fact that life's molecules are asymmetric with respect to their mirror images. For example, although most amino acids can exist in both left- and right-handed forms, life on Earth is made only of left-handed amino acids. Assuming that such a preference for "handedness" is characteristic of most life in the universe, the detection of chirality could reveal the abundant presence of living organisms on a planetary surface. To detect chirality, though, is a rather difficult task. It can perhaps be achieved through *polarization spectroscopy*. Light consists of oscillating electric and magnetic fields, with the two fields being always perpendicular to each other. By convention, the "polarization" of light refers to the direction of the electric field. When the fields oscillate in a single direction, we call it a linear polarization. In circular polarization, the fields rotate at a constant rate in a plane (either in the right-hand or left-hand direction) as the wave travels. Pigment absorption can produce a level of linear polarization—for example, it has been detected in Earthshine. The linear polarization is relatively high in visible light and low in near-infrared light, the opposite of the reflectivity of the Vegetation Red Edge, but scattering by dust can produce similar effects, so there is a danger of false positives. Circular polarization can on the one hand provide a more reliable signature of the optical action of amino acids, but on the other, the expected signal is extremely weak, and therefore probably impossible to detect.

Apart from stable gaseous and surface biosignatures, astronomers have also considered the possibility of *time-variable* biosignatures. Those include, for instance, changing concentrations of gases such as carbon dioxide, molecular oxygen, and methane, and variations in the reflectivity (albedo) of the planetary surface. On Earth such modulations are typically observed primarily as a result of the changing seasons and the associated variability in vegetation. However, since the expected variations are at best on the order of a few percent (and also prone to false positives), the reliable detection of such temporal biosignatures appears currently to be beyond the capabilities of even the next generation of telescopes.

We realize that this entire discussion of biosignatures had, by its very nature, to be somewhat technical. The overall picture that emerges, however, is quite simple and extremely exciting—we are either on the verge of discovering extrasolar life or at least we'll soon be able to place some meaningful constraints on how rare extraterrestrial life really is (in the case of non-detections). There is little doubt that the first signs of simple life to be detected on exoplanets are likely to be in the form of biosignature gases. Those gases could indicate a disequilibrium state in the atmosphere of an exoplanet, or could be produced in a biosphere that has experienced photosynthesis. Astronomers have realized that the detection of a high concentration of molecular oxygen (possibly through the discovery of ozone) would probably be the most promising sign in this respect. Until quite recently, however, researchers thought that the James Webb Space Telescope would not be able to detect oxygen. The reason was simple. JWST was not originally designed to scan distant planets for their oxygen concentrations, or, for that matter, even to look for the exoplanets most likely to be habitable. Rather, its original goal was to look deep into the cosmos, farther than the Hubble Space Telescope, and to show us the very first galaxies that formed in the universe. With its exclusively infrared vision, JWST was supposed to be blind to the most prominent spectral features of oxygen. But astronomers are quite resourceful, and used to never giving up when it comes to taking advantage of the availability of such a remarkable telescope. Consequently, as a first step, a team of researchers led by astrobiologist Joshua Krissansen-Totton from the University of Washington produced in 2018 an exciting theoretical result. They used computer simulations to show that the methane-plus-carbon-dioxide disequilibrium biosignature pair is plausibly detectable by JWST. Specifically, they found that JWST should be able to detect carbon dioxide and at the same time to constrain the methane abundance sufficiently well, so as to rule out known non-biological methane production scenarios (to about a 90 percent confidence level). Especially exciting in this respect was the fact that JWST could perhaps be able to achieve this feat in observations of the atmosphere of the most talked-about habitable-zone exoplanet—TRAPPIST-1e—which revolves in the habitable zone of a red dwarf with a mass of about 9 percent that of the Sun. TRAPPIST-1e is similar to Earth in mass and radius, in its surface temperature, and in the stellar flux it receives from its host star. This automatically made it one of the exoplanets most worthy of study when it came to the potential of habitability. Another system that receives much attention is TOI-700, which has two Earth-sized planets (TOI-700d and TOI-700e) orbiting their host star in the habitable zone.

Astronomers haven't totally given up on detecting oxygen either. In 2020, a group of astronomers led by Thomas Fauchez of the NASA Goddard Space Flight Center showed that JWST could possibly detect an oxygen feature at 6.4 microns, one which had previously not been extensively explored in exoplanet studies. Fauchez and his collaborators have shown that when oxygen molecules in the atmospheres of orbiting exoplanets collide (either with each other or with other gas molecules), the oxygen molecules can absorb infrared light at this particular wavelength, creating a strong absorption feature in the spectrum, which is detectable (in principle) in relatively nearby exoplanets. If JWST does indeed detect an oxygen-rich exoplanet atmosphere, this would be an extremely encouraging discovery, even though the presence of a single biosignature gas cannot be taken as an unambiguous proof of life.

JWST has already demonstrated its promising capabilities vis-à-vis determining the composition of an exoplanet's atmosphere. High-fidelity transmission spectra obtained by JWST in 2022 of the giant exoplanet WASP-39b detected carbon dioxide. This was the first time that carbon dioxide had been detected in an exoplanet. However, WASP-39b is definitely not expected to harbor life. It is a gas giant (with a mass of about 0.28 the mass of Jupiter), orbiting very close to its host star (an orbital period of just four days), making it extremely hot (about 1,650 degrees Fahrenheit, 900 degrees Celsius). It therefore belongs to a class known as "hot Jupiters." More impressively, in September 2023, a team led by Nikku Madhusudhan from the University of Cambridge used JWST to detect methane and carbon dioxide in the atmosphere of the sub-Neptune exoplanet K2-18b (with a radius of about 2.6 times that of Earth and a mass of about 8.6 Earth masses). The apparently abundant methane and carbon dioxide (both at a level of about 1 percent) combined with the non-detection of ammonia are at least consistent with some models suggesting the presence of a surface ocean underlying a hydrogen-rich atmosphere (with the ocean acting as a sink for the ammonia, basically sucking it out of the atmosphere). The methane itself is not likely to be produced by life, because in hydrogen-rich atmospheres it can be easily formed via photochemical reactions between carbon and hydrogen. Planets of the type of K2-18b have been dubbed *hycean planets* (a portmanteau of hydrogen and ocean). Most intriguingly, the researchers also found potential signs of dimethyl sulfide (DMS), although, by the time of this writing, not with a convincing statistical significance. If confirmed, this could be the first possible sign of extraterrestrial life, since DMS has been suggested to be a biomarker in hycean worlds. As we noted earlier in this chapter, on Earth DMS is produced, for example, by marine microalgae. We should note though, that in a study posted in January 2024, a team of researchers led by Nicholas Wogan of the NASA Ames Research Center suggested that the JWST observations of K2-18b can be explained by a gas-rich mini-Neptune with no habitable surface. In any case, these results definitely motivate future attempts to characterize the atmospheres of candidate hycean worlds. In particular, upcoming JWST observations should be able to confirm (or refute) whether DMS is indeed present in the atmosphere of K2-18b at significant levels.

To determine which kinds of atmospheric compositions will be considered as constituting a convincing proof of life on rocky planets, astrobiologist David Catling of the University of Washington and colleagues attempted to provide a probabilistic framework through which one can assess the likelihood of whether certain observational data represent a genuine detection rather than a false positive. Here are just a few illustrative examples of their results. They estimated that if we were to find an Earth-size planet in the habitable zone of its host star, and we could confirm the presence of a liquid-water ocean on its surface (say, via glint observations), and we also found that the planet's atmosphere was rich in molecular oxygen, and it contained nitrous oxide and methane too, the probability for this exoplanet to be inhabited would be 90 to 100 percent. For a similar exoplanet in the habitable zone, for which we would have detected only atmospheric molecular oxygen, accompanied by carbon dioxide and water vapor, the probability for it to be truly inhabited would drop to the range of about 66 to 100 percent. An exoplanet (still in the habitable zone) on which *only* molecular oxygen is detected, or *only* a combination of organic haze with methane in abundance is detected, or *only* a spectral feature indicating the Vegetation Red Edge is detected, would be about as likely as not to be inhabited. Even fewer such detections would lead to the conclusion that the planet is likely uninhabited.

We should keep in mind that determining the composition of the atmospheres of exoplanets is a serious challenge. Recall that even the mere detection of the exoplanets themselves was impossible until about three decades ago.

## Detecting Biosignatures

Currently, the most promising technique for detecting gaseous biosignatures is that of *transmission spectroscopy* (sometimes referred to as *transit spectroscopy*). This method was first suggested by astrophysicists Sara Seager of MIT and Dimitar Sasselov of Harvard, as soon as the first transiting exoplanet had been discovered. Very shortly after the theoretical idea was published, a group of astronomers led by David Charbonneau of Harvard successfully used transmission spectroscopy to identify for the first time the presence of a particular gas in the atmosphere of an exoplanet—the Jupiter-like exoplanet HD 209458b. The gas was atomic sodium (which is unrelated to life).

The idea behind transmission spectroscopy is genius in its simplicity. As an exoplanet transits its host star, some part of the star's light passes through the planet's upper atmosphere on its way to Earth. A fraction of that light is absorbed in the atmosphere, and the atmosphere's composition determines which wavelengths are absorbed and which are not. Consequently, by taking spectra when the planet is transiting, and when it is not, astronomers can obtain the transmission spectrum. These observational data can then be used (in combination with the known wavelengths at which various gaseous molecules absorb light) to discover which molecules are present in abundance in the atmosphere. Calculations of the expected strength of the signal show that employing this technique, one may be able to characterize the atmospheres of Earth-size planets around small M-dwarf stars, but not the atmospheres of similar exoplanets orbiting more luminous, Sun-like stars. In particular, astronomers hope to be able to detect water vapor (if it is present) in the atmospheres (if those exist) of exoplanets orbiting the M-dwarfs TRAPPIST-1 and Proxima Centauri (the closest star to the solar system).

Around Proxima Centauri astronomers have detected two planets (and a candidate), one of which is in the habitable zone. As we have already mentioned, the planet TRAPPIST-1e has been considered an especially attractive target, since not only is this exoplanet in the habitable zone of its host star (there are three such planets in the TRAPPIST-1 system), but this particular planet is also the most Earth-like in terms of its mass, size, and orbital position relative to the central star. If water vapor were indeed to be detected, this would suggest that TRAPPIST-1e has liquid water on its surface. Observations have already confirmed that TRAPPIST-1e lacks a cloud-free, hydrogen-dominated atmosphere, meaning that if it has an atmosphere at all, it is more likely to have a compact atmosphere like the terrestrial planets in the solar system. Another exoplanet that has received considerable notice is known as LP 890-9c or SPECULOOS-2c. It is a *super-Earth*—probably rocky and about a third larger than Earth. It orbits in the habitable zone of a red dwarf star that is about one hundred light-years away from us.

Results have already started coming in. In December 2022, astronomers presented preliminary results from JWST observations of the TRAPPIST-1 planetary system. In particular, astronomer Björn Benneke from the University of Montreal presented the first JWST studies of TRAPPIST-1g (the second-farthest planet from the central star). TRAPPIST-1g is the largest of the planets in that system, with a radius that is 1.154 times that of Earth. So far, the telescope has only been able to determine that the planet most probably does not have a primordial hydrogen-rich atmosphere—something that Cornell University astronomer Nikole Lewis and her team had previously already shown using the Hubble Space Telescope. Such an atmosphere would be physically puffed up owing to its very low density, so it would be relatively easy to detect. The fact that TRAPPIST-1g does not have such an extended atmosphere could mean that the planet has a denser, more compact, and more Earth-like atmosphere, composed perhaps of heavier molecules such as carbon dioxide, molecular nitrogen and water, or no atmosphere at all. Olivia Lim of the University of Montreal presented preliminary observations of the innermost planet—TRAPPIST-1b. Her observations also suggest that this planet, like TRAPPIST-1g, isn't surrounded by a puffy, hydrogen-rich atmosphere. Subsequent results by Thomas Greene of the NASA Ames Research Center and collaborators showed that TRAPPIST-1b is most likely a bare rock without any carbon dioxide in its atmosphere. Neither TRAPPIST-1b nor TRAPPIST-1g is in the habitable zone of TRAPPIST-1. In June 2023, another international team of researchers, led by Sebastian Zieba of the Max Planck Institute of Astronomy in Germany, used JWST to determine the amount of heat energy emitted by the rocky planet TRAPPIST-1c. The results suggested (still with some model-dependent uncertainties) that the atmosphere of this planet too, if it exists at all, is extremely thin. These findings were very interesting since, as co-author Laura Kreidberg from Max Planck pointed out, TRAPPIST-1c is, in some sense, a Venus twin: it's about the same size as Venus and receives a similar amount of radiation from its host star as Venus gets from the Sun. Therefore, Kreidberg noted, "We thought it could have a thick carbon dioxide atmosphere like Venus." Subsequent atmospheric modeling by a different team of researchers suggested that TRAPPIST-1c either had a relatively volatile-poor formation history (as compared to Earth and Venus), or lost a substantial amount of carbon dioxide during an early phase of hydrodynamic escape. Further investigations using advanced atmospheric simulations, however, raised possibilities of substantial oxygen or steam atmospheres. We should note that both TRAPPIST-1b and TRAPPIST-1c sit interior to what is known as the runaway greenhouse limit, where they could have potentially experienced radiation-driven complete atmospheric erosion. In contrast, TRAPPIST-1e and TRAPPIST-1f reside within the habitable zone, and likely experienced a comparatively brief steamy atmosphere during TRAPPIST-1's early life. Consequently, complete atmospheric erosion would perhaps not be expected in this case. Moreover, a study published in August 2023 by University of Bordeaux astronomer Franck Selsis and colleagues suggests that if a few of the TRAPPIST-1 planets were to partially transport heat internally by radiation, rather than fully by convection (fluid motion), then their surfaces would remain cool enough so that they might have been able to hold on to their water. It remains possible, of course, that all the TRAPPIST-1 planets formed without atmospheres. Furthermore, a study published in February 2024 suggested that the rapid orbital motion of TRAPPIST-1e could drive atmospheric heating (through electric currents) and thereby lead to a complete stripping of the atmosphere. We should note in this respect that the results from observations of the thermal emission from two other exoplanets, LHS 3844b and GJ 1252b, were also consistent with no atmosphere.

Other than transmission spectroscopy, astronomers can also use what is known as *occultation spectroscopy*—relying on the difference between the spectrum obtained when the "daytime" face of the exoplanet is fully visible, and when the exoplanet is totally eclipsed by its host star. However, since in this case the reduction in the observed flux depends not only on the size of the planet but also on its temperature (which determines its thermal emission), it turns out that occultation spectroscopy can be best achieved (if at all) in the infrared wavelength range of about eight to thirty microns. Unfortunately, the required sensitivity for such observations is somewhat higher than even that which JWST can reliably provide. There are a few other methods currently available for astronomers to use, but the chances of truly detecting biosignatures with those in the very near future are probably smaller than those which transmission spectroscopy offers.

The question is then what is the best way forward. The "Decadal Survey" released by the National Academy of Sciences in the US in November 2021 was unambiguous. Such a survey is carried out every ten years, to outline "Pathways to Discovery in Astronomy and Astrophysics" for the following decade. In its guidance and recommendations for astronomy and astrophysics, the decadal committee has given top priority to the search for extraterrestrial life. The committee called on NASA to start developing a space mission that will be able to determine whether extraterrestrial life in the Milky Way is absent, rare, or ubiquitous. To achieve this goal, the committee contemplated both *imaging* Earth-size planets in the habitable zones of their host stars and spectroscopic studies of those exoplanets for the characterization of their atmospheric composition. The plan is to first examine more than a hundred stars and their associated orbiting exoplanets, so as to identify some two dozen systems that can be followed up more purposefully. For the exoplanets in the selected group, the advocated mission would be expected to be able to detect such molecules as water, carbon dioxide, oxygen, and methane in their atmospheres—in other words, precisely the types of biosignatures that we have previously described as those that could provide the most compelling evidence. The proposed space telescope is supposed to be about six meters (twenty feet) in diameter and to be able to observe at a broad range of wavelengths from the ultraviolet, through visible light, to the infrared. Accordingly, the original concept was dubbed the Large Ultraviolet Optical Infrared Surveyor (LUVOIR). It should also be equipped with a coronagraph, a device that will effectively block out the light of the central star, thus allowing astronomers to image small, rocky exoplanets. The decadal survey steering committee estimated that such a telescope could be launched by the mid-2040s. The committee also recommended that at least one extremely large—around thirty meters (about a hundred feet) in diameter—ground-based telescope should be completed in the next decade. In the context of the search for life, such a telescope is expected to discover and study planets with masses as low as Earth's in the habitable zone, to obtain direct images of larger planets, and, via high-resolution spectroscopy, to characterize the atmospheres of transiting planets. The suite of instruments on this telescope will also allow astronomers to probe the earliest stages in the formation of planetary systems and to study protoplanetary disks around stars.

You must have noticed that most of the biosignatures we have discussed so far refer to life as we know it. This naturally raises the question of to what extent we can be sure that all life-forms in the universe share the same characteristics as life on Earth.

# CHAPTER 10

# Life as We Don't Know It

## The Design of Natural and Unnatural Life-Forms

*The unnatural, that too is natural.*

—Johann Wolfgang von Goethe

Chemical biologists have by now succeeded in showing that key components of life as we know it can all be generated from, hard to believe, cyanide—considered today to be a lethal poison—sulfur, and sunlight (UV), through a network of transformations that is progressively becoming better understood. This common framework suggests at least that life on our planet could have emerged naturally from the chemistry available on the early Earth. Suppose, however, that in some future exploration of our galaxy we find biological life with a different underlying chemistry. Would we be able to tell if that life had also popped up spontaneously? Which clues might hint that it could not have appeared through natural processes? In other words, we need to identify what to look for, to be able to distinguish natural life from artificial life that was the result of design by intelligent aliens. We should make it clear that at this point we are still discussing only an organic-type and not an electronic (artificial intelligence) form of life. We'll briefly discuss the possibility of intelligent machines at the end of this chapter.

Perhaps the best way to start thinking about the problem of life-as-we-don't-know-it and also how to recognize synthetic life would be to simply try ourselves to design and create new forms of life. This might seem like an audacious pursuit, especially since we have not yet been able to fully synthesize the kind of life we do know, with the type of chemistry that we are almost convinced can produce life (as we know it). What reason is there to even imagine that some different chemistry could lead to another variant of life? There is at least one clear rationale for being quite optimistic about the prospects of such a project: modern synthetic chemistry provides thousands of ways of making new chemicals, even through processes that could perhaps never occur in nature. This chemical repertoire is truly vast compared to the relatively small set of chemical variations on a theme that are currently used in the attempts to make the molecules of life on Earth. As a result, a sufficiently knowledgeable and imaginatively creative designer, whether human or alien, should be able, in principle at least, to come up with a multitude of alternative chemistries of life. In fact, chemists are presently exploring this nascent field, beginning with minor modifications to the chemistry that we believe led to life on Earth, and subsequently progressing stepwise to increasingly divergent ideas that may one day lead to very different life-forms. Here are just a few examples of these bold, fascinating endeavors.

Most of the molecules of life on Earth possess a distinct chirality. The nucleotide building blocks of DNA and RNA, and the amino acid building blocks of proteins are all one-handed, not superimposable with their mirror image, like our right and left hands. The amino acids characteristic of life are all "left-handed" in shape, while nucleic acids are "right-handed." Consequently, perhaps the simplest idea is to make a mirror-image version of life on Earth, in which every type of natural molecule would be replaced by its reverse facsimile. Indeed, a few daring scientists are actively trying to accomplish this goal. For instance, researchers in the laboratory of biologist Ting Zhu, now at Westlake University in Hangzhou, have used chemical methods to synthesize stretches of mirror-image DNA and RNA. They also used synthetic chemistry to make mirror-image protein enzymes and used those to replicate strands of DNA. That is, a mirror image of a protein characteristic of life on Earth can amplify a mirror-image version of normal DNA by mirror-image PCR. Moreover, the researchers subsequently synthesized a mirror-image protein that is an RNA polymerase enzyme so that they could transcribe their strands of mirror-image DNA into mirror-image RNA. The immediate goal of these experiments is the synthesis of a mirror-image ribosome that could be used to translate a mirror-image mRNA into mirror-image protein. If such experiments eventually lead to the ability to create mirror-image living cells in the lab, we would know with certainty that this life was the result of artificial design, since it would be distinct from any existing form of life on our planet. Still, the discovery of mirror-image life on any other Milky Way exoplanet would not be a huge surprise, since the particular handedness of life's molecules was almost certainly accidental, with either our existing chemistry or its mirror image equally likely to arise. We have previously mentioned that research published in 2023 suggests that the mineral magnetite is likely to have been the most suitable natural agent to accommodate a process for the generation of homochirality in the molecules of life on Earth. Notably, this process selects for either right- or left-handed chirality depending on the direction of the applied magnetic field, which would be opposite in the Northern and Southern Hemispheres. Thus the chirality of life's building blocks may come down to an ancient accident of place of birth.

In relation to the common cellular life we are familiar with, there is another intriguing question that comes up: whether it is possible to construct cell membranes from non-biological building blocks. Recall that membranes provide the important characteristic of compartmentalization to life. On the one hand, they create physical boundaries to cells, and on the other, they control molecular transport into and out of the cell. To investigate the possibility of unusual cell membranes, chemical biologist Neal Devaraj and his colleagues at the University of California, San Diego, are developing reactions that can trigger vesicle formation and reproduction. Specifically, they have taken advantage of a Nobel Prize–winning class of chemical reactions known as "click chemistry," which are often used to connect small units together to create more complex molecules. The Devaraj group uses this chemistry to join together two single-chain lipids to make a two-chain phospholipid that is very similar to the two-chain phospholipids found in modern biology. These synthetic phospholipids assemble into bilayer membranes just like natural phospholipids. However, the cellular process for phospholipid synthesis requires a series of complex enzymes, while the click-chemistry approach is so simple and fast that it does not require enzymes. What is perhaps most interesting about these experiments (and which may or may not apply to realistic prebiotic conditions on an exoplanet) is the fact that Devaraj's group has succeeded in developing a catalyst for this reaction that can even catalyze its own synthesis. That is, as long as the right synthetic building blocks are being provided, these artificially created membranes could grow and divide indefinitely, playing the role of a non-biological protocell membrane system.

The even more intriguing question is, of course, whether there can be modifications to the very central molecules of biology, DNA and RNA. This inquiry is further motivated by the fact that we know, by way of a few decades of aggressive synthesis experiments, that there are many variants of DNA and RNA that look as if they could serve as perfectly good genetic molecules. Could some of those be used as foundations for the design of artificial life? This may sound like science fiction, but Szostak's lab is determinedly trying to develop a synthetic genetic system, based on a variation to the structure of DNA, which would, in effect, even improve its ability to copy itself! Specifically, Szostak and his colleagues have been working with a molecule called *phosphoramidate DNA*, whose chemical structure is the same as that of DNA, except that an oxygen atom at a certain position in every sugar component has been replaced by a nitrogen atom. As a result, the modified nucleotide building blocks are much more reactive, and the molecules are easier to copy in this chemical system, without the help of enzymes. We should note that while Szostak's team has not succeeded yet in fully replicating short single strands of its synthetic DNA, scientists involved in the research do feel that the moment when they can construct living cells that use a genetic material different from standard RNA and DNA may be right around the corner. Consequently, if we were to find exoplanetary life that uses this, or another similar version of DNA, it would certainly be interesting, but based on our current knowledge we would not be able to fully rule out a natural origin for such life.

The examples we have discussed so far have still not wandered too far afield from the biology on Earth. This leads us to the question of whether life with an entirely different chemistry is possible. In other words, we would like to know whether we, another intelligent species, or nature could design something that is utterly distinct from the life we know. For researchers, this is an incredibly alluring, if taxing task, and a grand challenge for the coming century of chemistry. The discovery of numerous lakes of liquid methane and ethane on the surface of Saturn's moon Titan (described in Chapter 8) has also inspired research in this direction. Realizing, however, that it is very difficult to work with liquid methane in the laboratory, scientists have turned their attention to other nonpolar organic solvents, such as the hydrocarbon *decane* (composed of ten atoms of carbon and twenty-two of hydrogen). In fact, as early as 1991, chemist Hironobu Kunieda of the Yokohama National University in Japan and his colleagues succeeded in producing inside-out membrane vesicles in decane. By "inside out" we mean that unlike in the membranes of normal cells, in the Kunieda membranes the polar parts of the lipids were in the interior, while the hydrophobic (water-repelling) parts were sticking outward into the (also hydrophobic) solvent. Interestingly, in spite of being inside out and composed of very different molecular constituents, Kunieda's vesicles looked like the normal vesicles we find in life as we know it. This naturally raises the question of whether it is also possible to construct genetic polymers to replace RNA and DNA in such nonpolar solvents. While we don't know the answer to this question yet, a few researchers find the challenge involved in designing and producing such genetic material irresistible. Consequently and unsurprisingly, attempts in this direction are going full speed ahead.

This brings up again the possibility of cold life on Titan itself, perhaps happily swimming in those liquid methane/ethane lakes. As we have already noted in Chapter 8, however, as tantalizing as this idea may sound, liquid methane/ethane is such a poor solvent that it may be impossible to have any kind of complex chemistry taking place in such a harsh environment. For now, therefore, we have to admit that visions of such highly disparate life-forms on Titan may have to remain in the realm of science fiction.

Nevertheless, the idea that alternative life-forms may emerge in solvents other than water continues to intrigue researchers. University of Florida chemist Steven Benner and colleagues already discussed this possibility extensively in an article published in 2004. They concluded (albeit speculatively) that life "may exist in a wide range of environments. These include non-aqueous solvent systems at low temperatures."

Chief among other potential solvents that researchers have considered over the years is ammonia (composed of a nitrogen atom bonded to three hydrogen atoms). Indeed, given that ammonia (like water) is abundant in the cosmos, the possibility of extraterrestrial life using ammonia as a solvent was raised as early as 1954 by British geneticist J. B. S. Haldane. Like water, ammonia can dissolve many organic molecules. In addition, it can dissolve quite a few metals too. However, ammonia's ability to sufficiently concentrate prebiotic molecules and to hold them together so as to allow for the synthesis of self-reproducing systems remains questionable.

The list of other potential solvents includes, for instance, hydrogen sulfide (composed of two atoms of hydrogen bonded to an atom of sulfur), which is quite similar to water in its chemical properties (but is also known for its pungent rotten-eggs odor), and is abundant in areas of volcanic activity (such as Jupiter's moon Io). Still, what may be a serious disadvantage to hydrogen sulfide as a solvent for life is the fact that it is in liquid form only in a relatively narrow temperature range (from about minus 122 to minus 75 degrees Fahrenheit), although the range is somewhat wider for higher pressures.

The tentative detection of phosphine in the atmosphere of Venus (described in Chapter 7) has revived interest in the possibility that some form of life may exist in Venus's clouds. As might have been expected, since those clouds consist primarily of sulfuric acid, the question arose of whether sulfuric acid could also play the role of a solvent for life's molecules. The main problem with this idea is that sulfuric acid is a powerful dehydrating agent, which can remove water from the chemical structure of many compounds. For example, the addition of concentrated sulfuric acid to a pile of common sugar results in a smoking pile of blackened debris. Nevertheless, MIT biochemist William Bains and colleagues examined the question of whether life might be able to evolve and adapt to sulfuric acid as a solvent, in an article published in April 2021, using the clouds of Venus as a test case. Surprisingly, these researchers concluded that microbial life might be able to adapt to concentrated sulfuric acid as a solvent, but that more work needs to be carried out to explore whether, and how, life could truly *originate* in such a solvent.

Given that an information system is one of the crucial characteristics of Darwinian evolution, other groups of scientists have taken baby steps toward the design of new materials that can store and transmit information through replication. So far, the amount of information stored in the sequences of such novel polymers is very small, but this is an exciting venture, and the groundwork is being laid right now. At some future time, the design of alternatives to DNA may become so common as to metamorphose into a high school project, much as playing with and manipulating DNA and genome editing have been transformed from Nobel Prize–winning accomplishments to commonplace undertakings in schools and an attractive activity for biohackers.

The question then becomes whether we can go beyond chemical replacements for DNA and RNA as we strive to dream up alternatives to life as we know it. For example, life based on silicon instead of carbon is an old idea (first proposed in 1891 by German astronomer Julius Scheiner), which has been extensively discussed, because silicon is in the same group as carbon in the periodic table and it has somewhat similar chemical properties. At the same time, however, the rather limited diversity of silicon chemistry suggests to many that silicon-based biological life may be difficult or even impossible to construct (Carl Sagan referred to this notion as "carbon chauvinism"). MIT astrobiologists Janusz Petkowski, William Bains, and Sara Seager provided in 2020 a comprehensive assessment of the possibility of silicon-based biochemistry. Overall, they found that "the formation of many biological crucial functional groups is much less favorable for silicon than for their carbon counterparts." Consequently, they eventually concluded that "carbon chemistry is the chemistry of life, silicon chemistry is the chemistry of rocks."

In a very different type of study, a group of astrobiologists from Arizona State University examined in 2022 a more generalizable concept of biochemical universality. Using genomic databases, they investigated the enzymatic makeup of bacteria, archaea, and eukaryotes, thereby capturing most of Earth life's biochemistry. This allowed the researchers to identify statistical patterns in the biochemical function of enzymes that could be independent of Earth's component chemistry. In principle, these findings could provide some insights into the prospects of life originating in alien environments or for the design of synthetic life, but for the moment their practical implications are limited.

Where does all of this leave us? There is an extensive, unexplored chemical space that could potentially be the basis for the design of artificial, but still chemistry-based, life-forms, that could never emerge spontaneously in nature—not on Earth, not on any other planet. Producing such life-forms can be a thrilling scientific challenge, but a common worry is that we could lose control of such artificial life, which could go on to create havoc with our world. However, any initially designed artificial life-forms would be dependent upon chemically synthesized "nutrients" not found in the natural world, and therefore primitive artificial life would be confined to the laboratory in which it was created. Perhaps fortunately, the creation of artificial and chemically distinct life-forms that could live independently in the natural world is so much more difficult that the concept will remain in the realm of science fiction for the foreseeable future. If, with some far-future space mission, we were to discover such truly extraordinary life variants, we would know that those were artificially created and not natural.

There is, however, another branch of life as we (still) don't know it, one that is not based on biochemistry. The existence (or not) of that type of life depends on the answers to these questions: As some computer futurists suggest, will artificial intelligence become the dominant life-form here on Earth? And if so, is that the expected primary "intelligent" species in the Milky Way? Currently we cannot answer these questions. We do know, however, that although biological evolution on Earth has not stopped, it has already been far outpaced by cultural and technological evolution, leading to widespread speculation as to whether AI might replace us in the future.

In particular, one of the most intriguing unknowns in this equation is related to *consciousness*. Philosophers, psychologists, neuroscientists, and computer scientists have been vigorously debating whether consciousness is something that characterizes only the type of organic brains possessed by, for example, humans and other primates. In principle, it should not matter whether the electronics of a brain is based on biological neurons or designed silicon circuits. What may be more important is the "design" of the brain, the nature of its interconnections and ability to monitor and perceive itself, and the ways that it learns about the external world, including, crucially, other self-aware intelligent beings. However, at this point we simply don't know whether electronic intelligences, even if their intellects are superhuman in some aspects, such as memory and speed, would lack self-awareness. We have no idea if they will have a "true personality," or just the ability to emulate anything, or whether they will possess an inner life. Perhaps most crucially, we currently don't know whether consciousness is an *emergent* property, which any sufficiently advanced, sophisticated, and complex computer will eventually acquire. Computer scientist Edsger W. Dijkstra famously said that this question is irrelevant and semantic, like asking whether human-built submarines can swim. We must admit that we, the authors, don't find the question merely semantic. If machines can do nothing more than reproduce those aspects of human intelligence that are amenable to a mathematical/statistical treatment, leaving out many facets of human emotions and behavior, we would perhaps not accord their experiences the same value as we do our own. In that case, the post-human future would seem rather empty, and well, inhuman. If, on the other hand, computers can become truly conscious, we would have to accept the fact that their future hegemony here and elsewhere in the cosmos is a natural consequence of evolution in its broader sense. We should acknowledge though, that even if AI never acquires consciousness, it may dramatically influence the future cultural evolution of life. In the same way that social media are already having a significant impact on society, by dominating the meaning and use of language, AI may dictate in the future (at some level) the narrative of global communication.

There are many ways in which AI-based life would be different from life as we know it, even with regard to the methodology we should use to search for such life-forms. In Chapters 5 and 6 we saw that organic creatures most likely need a planetary surface environment, on which life can emerge from chemistry and subsequently evolve. But if post-humans make the transition to fully inorganic intelligence, they won't need either a "warm little pond" or an atmosphere in order to survive. They may even prefer a zero-gravity environment (meaning outer space), especially if they are interested in constructing massive artifacts. So searching for such intelligent beings on "habitable" exoplanets may be a waste of time. It may be in deep space that non-biological brains develop powers that humans can't even imagine.

Conceivably there are chemical, metabolic, and energetic limits to the size and processing power of organic brains. We may even be getting close to those limits already. The same limits, however, do not apply to or constrain electronic artificial intelligence machines, and are even less applicable, perhaps, to quantum computers. So, by any definition of what thinking, awareness, or comprehension entails, the capacity, intensity, and retrieval efficiency that can be achieved by organic, human-type brains will be undoubtedly swamped by the phenomenal cogitations that can be realized by AI-type machines.

The upshot of this short discussion is the recognition that the familiar dominance of a biological form of intelligence may be but a relatively brief, transient phase in the evolution of complex beings. The more permanent and unfamiliar stage may be one governed by creatures that emerge from the future of machine learning. We shall return to this topic in Chapter 12, when we discuss the search for intelligent civilizations.

# CHAPTER 11

# The Hunt for Intelligence

## Preliminary Thoughts

*The voice of the intellect is a soft one,*
*but it does not rest till it has gained a hearing.*
—Sigmund Freud, *The Future of an Illusion*

When it comes to extraterrestrial intelligent (in terms of their cognitive competence) and technologically capable (those who can create detectable signatures of advanced technology) species, our knowledge is on much shakier ground than when we discuss simple life-forms. The reasons are obvious. On the one hand, we know from astronomical observations of extremely distant galaxies that the laws of physics and chemistry are universal. There are also good reasons to think that some form of molecular Darwinism may be ubiquitous. On the other hand, we still cannot even estimate the odds of alien life *emerging* somewhere, let alone the chances of that life *evolving* to the status of an intelligent or technological civilization. We also have to admit that even if extraterrestrial intelligent civilizations do exist, we have absolutely no idea what the guiding motivations of such a civilization might be. Think of the variety of motives (e.g., existential, environmental, economic, ideological, political, nationalistic, egotistical, religious, and others) that have driven human endeavors in the past. As an extreme example of how different alien civilizations could be from our naive expectations, it has been suggested that they could be deeply contemplative and reclusive. This would certainly require an evolutionary history very different from our own. More speculatively, hyper-advanced AI beings may realize that it's easier to think and to undertake some computational operations at low temperatures, and therefore they may get as far away as possible from any star, to the most distant corners of the galaxy. At the opposite extreme on the spectrum of behavior, they could be expansionist and perhaps even more ferociously aggressive than we are, which seems to be the expectation of most of those who have given thought to the future trajectory of civilizations. Our ignorance is further compounded by the fact that we don't even know which of the many steps that have been apparently needed for the evolution of an intelligent species on Earth were truly essential, and which were just specific to the particular path that led to the development of human intelligence.

There are, nevertheless, several relevant things that we do know. First, any *detectable* intelligence must have reached a technological level higher than ours, since our efforts to date to search for extraterrestrial intelligence (SETI) could not have detected most of our own terrestrial transmissions. Second, based on how SETI searches are currently designed and conducted, to have a non-negligible chance of detection, the extraterrestrial civilization must have been already able to transmit for at least a few thousand years (otherwise the signal might not have reached us). This requirement has an additional important consequence. If we are to believe artificial intelligence experts, then the time span anticipated between the achievement of radio and laser transmission capability and the point at which AI starts to dominate over biological intelligence is not much longer than a few centuries (and possibly even shorter). If AI is truly to become the superior intelligence, then even if this estimate of the time it would take for AI to become the leading species is off by a thousand years, this still means that the bulk of intelligence in the cosmos (assuming it exists) may reside in machines, rather than in biological creatures. This realization should guide at least part of our SETI searches since even machines are likely to need energy sources and raw materials. They may also produce massive astro-engineering projects. Consequently, we should probably look for unnatural infrared sources that might betray waste heat that is virtually impossible to conceal, and we should also be alert to apparent "violations" of the laws of physics, since long-lived intelligent civilizations might have the ability to alter their cosmic environments in fundamentally unexpected ways. For instance, since a superior civilization may be able to control the weather on its home planet (assuming it continues to reside on the surface of a planet), it may produce bizarre cloud patterns in the planet's atmosphere. Moreover, unlike our search for simple biosignatures, we should definitely not restrict our hunt for intelligence only to habitable planets. Rather, vicinities of high energy sources (such as massive hot stars or perhaps even black holes) may be more promising.

In an attempt to quantitatively address the many uncertainties involved in estimating the number of civilizations capable of interstellar communication in the Milky Way, astronomer Frank Drake, a pioneer and spokesperson for the burgeoning searches for extraterrestrial intelligent life, formulated in the early 1960s a heuristic probability equation that bears his name. Even though this equation still cannot answer the main question for which it was originally formulated, it is worth examining a few of the developments that have taken place since its introduction. Sadly, Frank Drake himself passed away during the writing of this book, on September 2, 2022.

## The Drake Equation

The Drake equation is expressed as a product of a series of probabilities in the form (where the meaning of the various symbols is described below):

$$N = R_* f_p n_e f_l f_i f_c L$$

$N$ is the number we are trying to estimate—the expected number of civilizations in our galaxy with a technology that allows for the possibility of interstellar communication.
$R_*$ is the average rate at which new stars are being born in our galaxy.
$f_p$ is the fraction of those stars that have planets orbiting them.
$n_e$ is the average number of planets that can *potentially* allow for life to originate and evolve, for every star that hosts a planetary system.
$f_l$ is the fraction of those planets that could support life on which *in fact* life does arise.
$f_i$ is the fraction of those planets harboring life that succeed in evolving *intelligent* life.
$f_c$ is the fraction of intelligent civilizations that develop the necessary *technology* that enables them to engage in detectable interstellar communication.
$L$ is the length of *time* over which such civilizations release detectable interstellar signals.

The average star formation rate in the galaxy, $R_*$, has been known since the late 1960s. Estimates of the star formation rate rely mainly on such observations as the galactic UV emission from young stars and the tallying of the numbers of stars at different age ranges. Overall, these different methods consistently give a value of about *one to* *ten stars per year* for $R_*$. The main development in the last decade has been that observations performed with the Kepler space observatory together with other resources have provided quite reliable estimates for the next two factors in the Drake equation. The discovery of more than five thousand exoplanets by a variety of techniques (transits, radial velocity measurements, gravitational microlensing, imaging, and others, as described in Chapter 9), and statistical analyses based on those observations, have led to the conclusion that the fraction of stars that host planets, $f_p$, is of order *unity*. That is, almost all stars have planets orbiting them. Concerning the number of planets per planetary system that have conditions that could lead to the origin and evolution of life, $n_e$, astronomers typically take that to be the fraction of stars that have rocky (terrestrial) planets that are roughly Earth-size in their habitable zone. Recall that the habitable zone is that range of distances from the central star that allows for liquid water to exist on the surface of a rocky planet. Based primarily on Kepler data, this fraction is approximately (and conservatively) equal to 0.2, that is, about one of every five stars has such a planet. We should emphasize though that in terms of truly representing the planets that allow for life to emerge or be sustained, this value is associated with a number of uncertainties, which could move it either upward or downward. For example, we have seen in the case of a few of the moons of Jupiter and Saturn in the solar system that they possess large subsurface oceans that could, in principle, harbor life, even though these satellites are not in the Sun's habitable zone. This could increase the value of $n_e$. On the other hand, we have mentioned the fact that it is not clear whether habitable-zone planets around low-mass M-dwarf stars can truly support life, given detrimental effects such as stellar flares and tidal locking. This could decrease the average value of $n_e$.

We have no information on which to base any reliable estimates for the probabilities $f_l$ (the fraction of planets that could support life on which life in reality arises) and $f_i$ (the fraction of life-bearing planets that manage to evolve intelligent life), since we have not discovered yet any extraterrestrial life. In particular, whereas a few researchers have taken the fact that life on Earth emerged relatively rapidly (only a few hundred million years after Earth had cooled enough to become habitable) as evidence that abiogenesis (life arising from chemistry) must be common, a detailed statistical analysis has shown that one cannot reach such a conclusion on the basis of just one example of life. Although terrestrial life's early emergence does provide a clue that life *might* be abundant in the universe (if early-Earth-like conditions are common), this evidence is definitely inconclusive, and indeed is still statistically consistent even with an arbitrarily low intrinsic probability. On the other hand, it is true that discovering even a single case of life arising independently of the lineage of life on Earth (a "second genesis") would provide much stronger evidence that abiogenesis is not extremely rare in the cosmos.

As we shall see later in this chapter, a number of those who are skeptical about the existence of extraterrestrial intelligent life contend that the evolution of human intelligence has involved so many contingencies and improbable events that the chances of something similar repeating itself on exoplanets, $f_i$, is extremely low. At the same time, there are others who suggest that once life originates, the evolution to some form of intelligence is virtually inevitable.

The last two factors in the Drake equation, $f_c$ and $L$, are even less knowable. The first of these, which is supposed to represent the fraction of intelligent civilizations that achieve a technological level that allows them to produce detectable signals, is not unambiguously defined because it depends upon the detection sensitivity achieved by the receiving civilization. For a signal to be *detectable* requires that the civilization engaging in such a detection should itself reach a particular technological level. In this respect, we can only remark that if we were to transmit electromagnetic signals with our current absolutely best available technology, another nearby civilization at a similar technological stage could perhaps detect them, although only barely so.

The value of $L$, the timespan over which technological civilizations can generate detectable signals, is equally unknown. Various analyses of the Drake equation in the past have claimed, for various reasons, values typically ranging from about one thousand years to a hundred million years. The possibility of auto-extinction by nuclear war was perhaps the most powerful argument in favor of a low value. On the other hand, astrobiologist David Grinspoon argued, for example, that once a civilization has developed sufficiently, it might overcome all threats to its survival, thereby lasting for an indefinite period of time. Carl Sagan and Soviet astrophysicist Iosif Shklovsky argued optimistically for values of the order of a hundred million years. We, the authors, have to admit that other than noting the fact that humans have so far survived as a quasi-technological civilization for less than a century, we don't have a clue as to what the value of $L$ should be taken to be. Nevertheless, we should point out that the entire perspective on the value of $L$ may have been misguided. If, as artificial intelligence experts predict, non-biological machines will dominate intelligent civilizations in the future, then we must completely revise the concept of $L$, since an inorganic intelligence can in principle persist and continue to evolve, potentially for billions of years.

Given all of these uncertainties, just as an amusing exercise, we can attempt to guess certain values for the unknown factors, and plug those into the Drake equation. The great advances in astronomy in the past three decades allow us with some confidence to choose $R_*$ to be equal to about 10, $f_p$ to also be safely taken to be of the order of 1, and $n_e$ to conservatively be about 0.2. The remaining factors are nothing but not even particularly educated guesses. Let us assume that the fraction of planets on which life *does* arise (among those that *could* support life), $f_l$, is about 0.1. We shall be optimistic about the fraction of life-bearing planets that manage to evolve intelligent life, and assume that $f_i$ is also equal to 0.1 (even though quite a few researchers may be skeptical about such a high value; on Earth, for example, geochemical constraints delayed the appearance of large animals for four billion years). We feel slightly safer in assuming that if a species achieves intelligence, it can also reach the necessary technological level to allow it to develop interstellar communication (although some of our ancestors, such as *Homo habilis*, *Homo erectus*, and the Neanderthals, did not develop such a technology before they became extinct). We shall therefore (very optimistically) take $f_c$ to be equal to 1. As we have noted above, we have no clue concerning the value of $L$. Ignoring for the moment the possibility of non-biological AI, let's examine what we obtain if we use the minimum value of a thousand years and the maximum value of a hundred million years. Putting all of these numbers together gives us, for the expected number of civilizations in the Milky Way possessing a technology that allows for interstellar communication,

$$N(\text{min}) = 10 \times 1 \times 0.2 \times 0.1 \times 0.1 \times 1 \times 1{,}000 = 20,$$
$$N(\text{max}) = 10 \times 1 \times 0.2 \times 0.1 \times 0.1 \times 1 \times 100{,}000{,}000 = 2{,}000{,}000$$

Note that given the huge uncertainties in the values of at least three of these factors, this number could be easily reduced to 1—meaning that we are alone in the galaxy—or increased to many millions—implying that we are members of a quite large galactic community. The Drake equation, in this sense, doesn't achieve much beyond simply highlighting our ignorance. Of course, if we do accept the prediction that galactic technological civilizations are in fact dominated by non-biological machines (which can easily exist for billions of years), then $N$ can reach even higher values.

While there have been attempts to modify/improve the input into the equation in a variety ways, for example, by arguing that the fraction of habitable planets on which life truly emerges may not be constant, but rather a function of time (since, for instance, in the very early universe the abundance of the heavy elements forming terrestrial planets was very low), all of these additional subtleties can't overcome the fundamental absence of reliable observational data.

In spite of its obvious shortcomings, the Drake equation has in recent years been used to reach two interesting conclusions. One was pointed out in 2016 by University of Rochester astrophysicist Adam Frank and University of Washington astronomer Woodruff Sullivan. These researchers asked themselves the intriguing question: How many technological civilizations have *ever* existed throughout the entire history of the universe? Asking the question in this way, rather than inquiring whether those civilizations overlap in time with ours, allowed them to remove from the equation the highly uncertain value of the average lifetime of such civilizations, $L$. In addition, in their attempt to answer this alternative question, Frank and Sullivan judiciously combined the product of the first three factors of the Drake equation (those which have become known through astronomical observations) into one that reads:

$$\boldsymbol{f} = \boldsymbol{M} f_p \, n_e,$$

where $\boldsymbol{M}$ is the estimated total number of stars in the universe. Similarly, they fused the product of the remaining (unknown) three factors also into one:

$$\boldsymbol{U} = f_l \, f_i \, f_c.$$

Consequently, the revised equation for the total number $\boldsymbol{T}$ of technological civilizations to have ever existed now took the very simple form $\boldsymbol{T} = \boldsymbol{fU}$, where $\boldsymbol{f}$ is known from astronomy, and $\boldsymbol{U}$ encapsulates everything we don't know in the Drake equation. In other words, it measures the probability that life emerges and evolves to become an intelligent, technological civilization—what one might call the "bio-technological" probability. Quantitatively, we know from astronomical observations that there are approximately a hundred billion stars in a typical galaxy, and we also know from Hubble observations (and from the initial data from JWST) that there are no fewer than in the order of a trillion galaxies in the observable universe. Putting all of this together, the conclusion is simple (taking as before $f_p = 1$, and $n_e = 0.2$): unless the bio-technological probability is smaller than about one in 20 billion trillion ($2 \times 10^{22}$ in scientific notation) *humans would not be the only technological civilization to have ever arisen in the entire universe!* Of course, since we don't have any idea as to what the value of the bio-technological probability might be, we cannot really conclude from this fact that other technological civilizations *must* have existed. This simple exercise does, however, reveal how small this probability has to be for us to have been alone throughout the entire 13.8 billion years of the universe's existence.

The other interesting offshoot from Drake's equation was formulated by MIT astrophysicist Sara Seager. She composed in 2013 an equation similar in concept to Drake's equation, to estimate the number $N$ of planets with detectable signs of life in the form of biosignature gases. This alternative equation reads $N = N_* F_Q F_{HZ} F_O F_L F_S$, where the different factors on the right-hand side represent, in sequence: the number of stars that a certain astronomical survey observes; the fraction of those stars suitable for planet finding (the subscript $Q$ stands for "quiet" stars that don't obscure the search by being variable); the fraction of this subset of stars that have rocky planets in their habitable zone; the fraction of those planets that can be currently observed; the fraction of those that harbor life (of any form); and the fraction of those life-bearing planets in which detectable biosignature gases are produced.

Given the ongoing intensive efforts to detect life, Seager also attempted to put some numbers into this equation for a search with present instruments. She used estimates based on the expectations of being able to detect planets using observations with the Transiting Exoplanet Survey Satellite (TESS), and to search for biosignatures with JWST. The number of suitable stars from the TESS survey was estimated to be about

30,000. Out of those, about 60 percent were expected not to vary too much in brightness, so that planets orbiting them could be detected via the transit method. As we have noted above, the fraction of rocky planets in the habitable zone is (conservatively) known to be about 0.2. The fraction of planets that transit their host star (and therefore could be discovered by TESS) is about 0.1 for M-dwarf stars, and only for about 1 percent of those cases can the planet's atmosphere be observed in detail and be at least partially characterized by JWST.

Multiplying these numbers together, we initially find $N = 3.6\ F_L\ F_S$. The last two factors on the right-hand side are as unknown in Seager's version as were similar factors in the original Drake equation. If we adopt for $F_L$ (the fraction of planets that harbor life) the same optimistic value that we have adopted for the Drake equation (0.1), and we also assume (again optimistically) that the fraction of those planets harboring life that produce a biosignature detectable via transit spectroscopy is 0.1, then we obtain for the number of planets for which the TESS/JWST combination can detect signs of life the disappointingly low number of 0.036! This simple exercise exemplifies the magnitude of the problem of discovering biosignatures. Even employing the powerful TESS/JWST system, we'll have to get really lucky. The good news is, however, that with the next generation of telescopes (both in space and on the ground) the chances of detection will significantly improve.

Returning to the topic of extraterrestrial technological civilizations, as we noted in Chapter 1, the fact that we haven't yet seen any sign of the existence of such an advanced civilization in the Milky Way came as a surprise to the famous physicist Enrico Fermi. His puzzlement has become known as the *Fermi Paradox*.

## Where Are They?

Enrico Fermi may have been the last physicist capable of doing both highly original, complex theoretical work and groundbreaking experimental work. In 1926, he discovered the statistical laws (now known as "Fermi-Dirac statistics") that apply to particles such as electrons and neutrinos that are now collectively referred to as "fermions." Upon the discovery of nuclear fission—the reaction in which a heavy atomic nucleus splits into two lighter nuclei—he immediately realized that neutrons may be emitted in the process and that those could create a chain reaction. Consequently, he started directing a classical series of experiments in makeshift laboratories constructed on a squash court situated underneath the grandstands of the University of Chicago's Stagg Field stadium. This endeavor led quickly to the first controlled nuclear chain reaction. For his work on artificial radioactivity, and on nuclear reactions brought about by slow neutrons, he was awarded the 1938 Nobel Prize in Physics.

As we have briefly noted in the introductory chapter, the story related to the Fermi Paradox occurred during Fermi's summer visit to the Los Alamos Scientific Laboratory in 1950. The setting apparently was a luncheon with physicist colleagues Emil Konopinski, Edward Teller, and Herbert York. Konopinski later recalled that as they were walking to lunch, the four physicists exchanged humorous remarks about a *New Yorker* cartoon depicting aliens stealing public trash cans from the streets of New York. Later, in the middle of lunch, Fermi suddenly returned to the topic of aliens by asking another version of the titular question of this section: "Where is everybody?" He was expressing his surprise over the absence of any signs of the existence of other intelligent civilizations in the Milky Way galaxy. This question led to the concept of the Fermi Paradox. As an amusing aside, we should note that naming this problem after Fermi is just another example of a phenomenon that has become known as "Stigler's law of eponymy," which states that no scientific discovery is named after its original discoverer. In fact, Russian visionary rocket scientist Konstantin Tsiolkovsky had discussed the issue of the absence of evidence for the existence of advanced extraterrestrial civilizations in a short article, which he had already published in 1933. Moreover, Tsiolkovsky even thought that he knew the solution to this paradox—in his opinion, advanced aliens must have considered humans not mature enough for an encounter or even a contact.

Returning to the paradox itself, on the face of it, Fermi's bewilderment was justified. Even using the types of chemically driven rockets that humanity currently possesses (but also utilizing the assistance of what is known as a *gravitational slingshot*—using the gravity of astronomical objects to change the direction of and to accelerate a spacecraft), one can reach even the most remote corners of our galaxy in a time that is about ten to a hundred times shorter than the age of the galaxy. Moreover, the puzzle of not having seen so far any hints of advanced civilizations becomes even more pronounced and astonishing when we consider the fact that presumably such civilizations would not have remained stuck with the technology of our snail-paced (relatively speaking) rockets, but rather could have achieved space travel at a significant fraction of the speed of light. And that is not all. The mystery is further exacerbated if we assume that some of these interstellar probes could, in principle at least, be self-replicating.

Nobody really knows the solution to Fermi's Paradox, but over the years that have passed since Fermi asked the question, more than a hundred potential solutions have been suggested. In some sense, the easiest solution is, of course, to think that we are truly alone in the Milky Way and no other technological intelligence exists, or that such civilizations are so exceedingly rare and far apart that the chances for any type of contact are minuscule.

One of the relatively early skeptics about the existence of alien intelligent civilizations was cosmologist Brandon Carter, a distinguished theoretical physicist known for his work on black holes, and for formulating the *Anthropic Principle*—the notion that what we can expect to observe about the current structure of the universe, the values taken by the constants of nature, and the laws of physics, have to be restricted to the conditions necessary for our own existence as observers. As long as four decades ago, Carter already raised doubts about the existence of extrasolar complex or intelligent life. His argument was based on the seemingly remarkable coincidence concerning the time it has taken life to appear and evolve to intelligence on Earth (about four billion years) and the "window of opportunity" that the evolution of the Sun offers for life on Earth to exist. Carter estimated that this window extends from about 3.8 billion years ago, when the bombardment of Earth by asteroids started to abate, to about 800 million years hence, when the Sun will become so hot, on its way to becoming a red giant, that it will sterilize the planet. Consequently, unless life on Earth had managed to emerge relatively quickly (astronomically speaking), humans would not have had the chance to evolve before the habitability window of opportunity closed.

We should note that on the face of it, the coincidence (to within a factor of two) between the two timescales—basically the lifetime of the host star (e.g., our sun) and the time to evolve an intelligent species on a habitable planet (e.g., the Earth)—seems indeed surprising, since a priori these timescales appear to be completely independent quantities. The time to evolve intelligence is presumably determined by chemical and biological processes on a planet's surface, while the lifetime of a star is dictated by the energy available from nuclear fusion reactions at the star's core, the star's luminosity, and the interplay between those and the force of gravity. From this unexpected coincidence in the case of the Earth-Sun system, Carter claimed to have shown that the time required to evolve an intelligent species would *typically* (meaning, almost always) be much longer than the window of opportunity that stars afford their habitable planets, and therefore, that intelligent life would be exceedingly rare. Earth, according to Carter's argument, had to be one of those very few and far between exceptions.

Here is a very brief explanation of Carter's argument in non-mathematical terms. Let's examine two timescales: the timescale for the evolution of intelligent life on the surface of an exoplanet, and the lifetime of the host star of that planetary system. If these two timescales are truly completely independent, then the probability of the two of them being of the same order of magnitude is extremely small. Rather, for two entirely independent quantities, each one of which can assume a very broad range of values, it is much more likely that one of them would be much larger than the other. However, if *generally*, the time to evolve intelligent life is much shorter than the lifetime of the star, it would be extremely difficult to understand why *in the very first system* in which we have encountered intelligent life—the Earth-Sun system—we found the two timescales to be nearly equal (to within a factor of two). Rather, we should have found that evolution had produced an intelligent species when Earth was much younger (say, only one hundred million years old). On the other hand, if *generally*, the timescale to evolve intelligent life is much longer than the lifetime of the star—meaning that *generally* intelligent life does not develop in the afforded window of opportunity—we would in fact *expect* that in the first system where we encounter intelligent life, the two timescales would be nearly equal, since only in the extremely rare cases in which the evolutionary time was just about equal to the star's lifetime, would intelligence have had the chance to appear (since the time to evolve intelligence has always to be shorter than or equal to the lifetime of the star). From the near-equality of the two timescales in the Earth-Sun system, therefore, Carter concluded that *typically* intelligent civilizations do not have sufficient time to evolve, and Earth is a very rare outlier.

In an article published in 1999, one of us (Livio) attempted to scrutinize Carter's argument. He noted that if one could show that as the lifetime of the star increases, so does the evolutionary time, then we would even *expect* to observe that the two timescales would be nearly equal in the first system that we find to harbor complex life. The reason is simple. The evolutionary time cannot exceed the lifetime of the star since life needs the star as a source of energy. At the same time, we know that the number of stars increases with increasing lifetime (there are many more low-mass stars, which also live longer). Livio therefore used a simple "toy model" to show that the two timescales could (in principle at least) be related, rather than being entirely independent. For example, in the atmospheres of planets, ultraviolet (UV) radiation from the star can lead to the initial rise of oxygen (and ozone) in the planet's atmosphere, by splitting water molecules. Because of its absorption properties, the ozone is necessary to protect emerging multicellular land life from UV radiation. Even more generally, as we have seen in Chapter 3, the flux of UV light plays an important role in the processes related to the origin of life. Crucially, the intensity and spectrum of UV light that a star emits is determined by the star's type (its surface temperature and luminosity)—the same characteristics that also determine its lifetime. Livio therefore suggested that Carter's gloomy conclusion about the nonexistence of extraterrestrial intelligent civilizations may not be altogether unavoidable. Moreover, the long time it took for oxygen to rise to protective levels in Earth's atmosphere, and coincidentally to levels that could support the energy requirements of large animals, could also provide a simple explanation for the delayed arrival of complex life. This delay (by some three to four billions of years) was taken by British zoologist Matthew Cobb to mean that there is no significant evolutionary drive to complex multicellularity. Like Carter, Cobb therefore also concluded that intelligent life must be exceedingly rare. Contrary to this opinion, the requirement for an oxygen-rich atmosphere indicates that complex life may, in fact, have appeared on Earth rather quickly, but only once the conditions of high oxygen levels allowed for it to emerge.

Astrophysicist Milan Ćirković of the Astronomical Observatory of Belgrade and colleagues raised another, more basic objection, to Carter's argument. They pointed out that because of a variety of astrophysical and planetary phenomena that are potentially catastrophic to life (such as gamma-ray bursts, supernova explosions, freezing "snowball" episodes, etc.), the two timescales used by Carter for his reasoning (the lifetime of the star and the time to evolve an intelligent civilization), aren't particularly well-defined, thus significantly weakening his analysis.

We have already mentioned that another skeptic of the existence of extrasolar life is cosmologist, astrobiologist, and author Paul Davies. His pessimism starts with his conviction that the fundamental laws of physics and chemistry are "life blind." That is, there is nothing in those laws, in his view, which singles out *life* as a favored final state or an ultimate goal. What's more, he claims that he has not even seen "a convincing theoretical argument for a universal principle of increasing organized complexity." In other words, Davies suggests that there is nothing to prevent a soup of messy chemicals from remaining chaotic forever (although as we have seen this is not how life began!). In addition, in an attempt to attach a more specific, probabilistic weight to his pessimism, Davies presented the following argument: Suppose that life's origin needs a particular sequence of ten critical and precise chemical steps (he assumes that ten steps represent, if anything, an underestimate to the actual number of critical steps that are truly required). Assume further that each one of those steps has a probability of occurrence (during the period throughout which an exoplanet remains habitable) of 1 percent (again, a value he considers to be optimistic). Then the combined probability for life to originate is the staggeringly low one in a hundred billion billion. In this case, even with an estimated presence of perhaps as many as hundreds of millions of habitable planets in the Milky Way, the odds of there being a second planet harboring life in our galaxy (other than Earth) would be negligible. Davies's bleak conclusion: "We are probably the only intelligent beings in the observable universe, and I would not be very surprised if the solar system contains the only life in the observable universe."

We, the authors, think that the current research on prebiotic chemistry makes such conclusions premature. As we have described in Chapters 2–5, the most recent findings by origin-of-life researchers seem to suggest that the entire outlook on the origin-of-life question for the past four decades was misguided. The step-by-step "which came first" debate originated from a scenario that assumed that one must find a way to construct the first cells one subsystem (e.g., informational, metabolic, compartmentalization) at a time, with each subsystem paving the way for the next one. This has changed dramatically in the last few years. Current thinking and experimental results suggest that one could produce the building blocks for *all* the subsystems all at once. Laboratory research hints that in spite of being very intricate entities, the first cells might have emerged from a mixture of chemicals that was dominated by the necessary building blocks, and not from the staggeringly complex mixtures generated by, for example, the Miller-Urey experiment. Accordingly, rather than assuming a one-step-at-a-time process, what researchers are now trying to do is to draw a picture of a complete, robust pathway to life, which would integrate all the data from studies of prebiotic chemistry with observations from geology, atmospheric chemistry, and astrophysics. This is not to say that we know the probability for the emergence of biology from chemistry. That probability may still be very small, but it is premature to assume that we can estimate its value.

Physicist and author John Gribbin agrees with Davies on the scarcity of technological civilizations. He wrote a book entitled (rather depressingly from the perspective of the present authors) *Alone in the Universe: Why Our Planet Is Unique*, in which he listed a series of properties of the solar system (such as, the characteristics of the Sun itself; the presence of the giant planet Jupiter; etc.) and a string of cosmic events in the history of Earth (such as the formation of the Moon as a result of a Mars-size planet colliding with Earth; the asteroid impact that brought about the demise of the dinosaurs; etc.), which he thinks were ultimately responsible for a unique form of intelligent life emerging on our planet. He closed his book with the categorical statement: "The reasons why we are here form a chain so improbable that the chance of any other technological civilization existing in the Milky

Way galaxy at the present time is vanishingly small. We are alone, and we had better get used to the idea."

Both Davies and Gribbin expressed their reasoned beliefs in the paucity of intelligent life in the galaxy about a decade after geologist Peter Ward and astronomer Donald Brownlee published their (somewhat controversial) well-researched and systematic book *Rare Earth: Why Complex Life Is Uncommon in the Universe.* Ward and Brownlee presented a long list of criteria ranging from the existence of plate tectonics to the planet being accompanied by a relatively large moon, and an enumeration of unlikely contingencies, all of which in their view had played a crucial role in the evolution of complex life on Earth. To them, the bottom line was simple: microbial life may be common in the universe, but complex and intelligent life is likely to be extremely rare. We must say though, that we, the authors, suspect that Ward and Brownlee may have started from their conclusion and went in search of arguments favoring their bias.

We can also attempt to examine the question of whether or not technological civilizations may indeed be scarce in our galaxy from the perspective of the Drake equation. Going in the order of the respective components of the equation, a dearth of technological species may result from any one of the values of the last five factors $n_e f_l f_i f_c L$ (or more than one) being extremely small. Just as one example, we mentioned in Chapter 6 that a few studies suggest that asteroid impacts on the early Earth may have been important for the emergence of life. The main reason is that the interaction of the asteroids' hot iron core with water can speed up the stockpiling of hydrogen cyanide, which, as we have seen in Chapter 3, was crucial for prebiotic chemistry on Earth in lakes and their sediments (mostly in the form of ferrocyanide). Since the generation of a stable asteroid belt, and the driving mechanism directing asteroids from that belt to impacting Earth required (in the solar system) the presence of both planets Jupiter and Saturn, the absence of a similar giant-planet architecture in other solar systems could (in principle) result in a lower probability for the emergence of life. We should make it clear though that this is only an example. We don't know with certainty whether asteroid impacts are truly a necessary condition for life. Moreover, the Earth's formation period must have included a phase of gradually declining bombardment with relatively large impactors.

Overall, however, it does seem to us, the authors, somewhat more likely that if it is indeed the case that we are basically alone in the Milky Way, this would not be merely because life has completely failed to *emerge* everywhere else, but also because the evolution to an intelligent, technological species, capable of producing detectable signals, does genuinely appear to involve a series of rather low-probability steps. Suffice it to note, for instance, that it is not even certain whether a technological civilization would have emerged on Earth, were it not for that fateful asteroid impact, some sixty-six million years ago, which wiped out the dinosaurs (together with 80 percent of all animal species). Moreover, history has indeed shown that even earlier lineages of hominins, the ancestors of *Homo sapiens*, had become extinct long before achieving any technology more advanced than a few very basic, primitive tools.

Another suggested solution to the Fermi Paradox, still in the general "we are alone" category, is the assumption that the *lifetime* of technological civilizations, represented by the factor $L$ in the Drake equation, is rather short. This could be the result of either civilization-produced catastrophes (such as a nuclear war), or of natural cosmic risks. However, we find such solutions to the paradox implausible. While there is no denial that a nuclear war or some biotechnological disaster (such as one causing a runaway pandemic) could wipe out our current civilization, it may be less likely that such catastrophes would lead to the extinction of our species. Furthermore, it is hard to believe that this would be the fate of *all* advanced galactic civilizations. Similarly, while the results of an impact by a large asteroid or the explosion of a nearby supernova could be devastating for life, presumably exceptionally progressive civilizations (if they exist) would have developed adequate risk-mitigation and defense mechanisms against such disasters. To prove this point, even NASA's Double Asteroid Redirection Test (DART) investigation in September 2022 demonstrated that the kinetic impact of a spacecraft the size of a golf cart with a target asteroid about the size of a baseball field (called *Dimorphos*) could successfully alter the asteroid's orbit. This marked the first time, by the way, that humans purposely changed the motion of a celestial object and the first full-scale demonstration of asteroid deflection technology.

There is another, more philosophical objection that we, the authors, have to proposed solutions of the Fermi Paradox that invoke the "we are alone" route. The fact that we are scientists does not mean that we are altogether free of, or immune to, personal opinions and emotions. We feel that to think that the only place in the galaxy where life exists is Earth is nothing short of arrogant, and far too anthropocentric. Such a belief, without any convincing experimental or observational evidence, perhaps had its place in the past history of the human psyche up to the scientific revolution, but not after it. We are more inclined now to adopt a humility that has become known as the *Copernican Principle*.

Astronomer Nicolaus Copernicus proposed in the sixteenth century that the Earth on which we live is not at the center of the solar system. In case you wonder, there is a compelling reason as to why the Copernican model, which was originally suggested simply to explain observations of the solar system, has since been elevated to the status of a *Principle*. In the centuries that have passed since Copernicus's idea was published, it seems that this humbling concept—that on the cosmic scale humans are nothing special—has continuously gained further support through a series of subsequent monumental discoveries.

First, Charles Darwin showed that rather than being the apogee of creation, humans are simply an ordinary product of evolution by means of natural selection. Second, astronomer Harlow Shapley showed at the beginning of the twentieth century that the solar system itself is not at the center of the Milky Way galaxy. It is in fact almost two-thirds of the way out, in the galactic suburbs. Third, as we have already noted, recent estimates based on searches for extrasolar planets in the Milky Way put the number of Earth-size planets in that "Goldilocks" habitable zone of their host stars potentially in the hundreds of millions, if not more. Fourth, astronomer Edwin Hubble (after whom the Hubble Space Telescope is named) showed already in 1924 that there exist galaxies other than the Milky Way. Moreover, the most recent estimates of the number of galaxies in the observable universe give the staggering number of about two trillion. Even the stuff we are made of—ordinary (*baryonic*) matter—appears to constitute less than 5 percent of the cosmic energy budget, with the rest apparently being in the form of *dark matter*—material that does not emit or absorb light—and *dark energy*—a smooth form of energy that permeates all space. And if all of this is not enough, in recent years, many theoretical physicists have started to speculate that even our entire universe may be but one member of a huge ensemble of universes—a *multiverse*.

The Copernican Principle suggests therefore that we should exercise humility—that from a purely physical perspective, we are just a speck of dust in the grand cosmic scheme—rather than believe that we are so privileged that we are the only intelligent technological civilization in our galaxy. Consequently, we, the authors, are not prepared (at least not yet) to accept the "we are alone" solution, unless some concrete, compelling evidence emerges to suggest that other technological species indeed do not exist.

Two other classes of ideas have been proposed to resolve the Fermi Paradox. One broad category contends that extraterrestrial civilizations have in fact already been here in the past, or, believe it or not, even that they are here right now. While to date there is absolutely no unambiguous evidence to support such a hypothesis, it would be difficult to refute several of the presented scenarios. For example, a few scientists, including astrophysicist Fred Hoyle, and biologists Francis Crick and Leslie Orgel, have argued for *panspermia*—microbes and extremophiles migrating from space—as being the source of life on Earth. Panspermia is a rather unsatisfying hypothesis, as it simply pushes back the origin of life to some unknown earlier planetary environment. The reason that panspermia could be relevant to the Fermi Paradox is that Crick and Orgel speculatively suggested that the panspermia could have been deliberate, meaning that life on Earth was planted by some alien advanced civilization. One way to test this class of potential solutions would be to search more vigorously for the possibility of alien artifacts lurking within the solar system. We shall mention a few attempts in this direction later.

A second scheme is even more extreme. It goes along the lines imaginatively depicted in the dystopian science fiction movie series *The Matrix*. In other words, it submits that humanity, and perhaps even the entire cosmic vistas we observe, are unknowingly trapped inside a simulated reality designed and run by some hyper-intelligent civilization. While there is no way in which we could prove such a hypothesis wrong, accepting it as correct does not aid us in deciphering the origin of life any more than believing in panspermia or a supernatural origin of life.

The last group of proposed solutions to the Fermi Paradox assumes that technological civilizations do exist, but that humanity has not yet been able to discover them. This presumed failure to detect any technosignatures could be either because these civilizations are really not interested in advertising their existence (or maybe they even deliberately prevent their detection), or because our technology or methodology is inadequate and not yet at a sufficiently high level to allow for successful detections. We could be searching in the wrong places (for instance, artificial intelligence exo-civilizations may not be associated with planets), we could be using the wrong methods (for example, these aliens may not be using what for them could be long-antiquated radio communication), or we are simply unable to identify and decrypt the signals of a far superior civilization. Of the three broad categories of suggested solutions to the Fermi Paradox ("we are alone"; "they are here"; "they exist but we haven't found them yet"), this third class seems to us to be perhaps the most plausible. Since technologies, once emerging, develop rapidly, if technological civilizations are not really ubiquitous, the chances of such a civilization being at a synchronized evolutionary stage with ours are relatively very small. Rather, such civilizations are more likely to be separated, in terms of their technological prowess, by hundreds of millions or maybe even billions of years. In that case, given our relative youth as a technological species, we can be expected to be the one that is far behind. If indeed, as the AI gurus suggest, after a relatively brief phase of biological intelligence, civilizations become dominated by intelligent machines, then even the premise at the basis of the Fermi Paradox may be wrong. The paradox stemmed from the assumption that intelligent civilizations are expansionist, and they strive to reach every corner of our galaxy. But if intelligence in the galaxy is truly dominated by thinking machines, this can, on one hand, completely change the implications of the Drake equation, and on the other, it may offer a new solution to the Fermi Paradox. Concerning the Drake equation, while flesh and blood civilizations may be short-lived, the machines can be nearly immortal. Second, since the machines have not evolved by means of natural selection (which favors and puts a premium on "survival of the fittest"), they may not be driven by the same types of aggressive tendencies that characterize humans. For instance, they may not be seeking at all to colonize the galaxy, and hence Fermi's basic argument for their nonexistence because we haven't seen them yet, would not apply. These far superior species may rather be leading studiously introspective lives. For all we know, they may be content with staying at their home locations, and contemplating how to improve their world.

What would be the consequences of a successful detection of a technosignature? The sequelae would naturally be quite different depending on the suspected source. For example, a signal that was pinpointed as originating from a particular exoplanet would provide us with a physical location of an alien civilization, with all the associated ramifications of such a discovery (e.g., the possibility of communication). A signal not associated with any planet, and encoded beyond comprehension, on the other hand, may only have more philosophical, rather than immediate practical implications (e.g., for our perceived self-worth or for religious beliefs).

# CHAPTER 12

# The Hunt for Intelligence

## The Searches

*I, for one, am not so immensely impressed by the success we are making of our civilization here that I am prepared to think we are the only spot in this immense universe which contains living, thinking creatures.*

—Winston Churchill, "Are We Alone in the Universe?"

In 2016, while one of us (Livio) was on a visit to the US National Churchill Museum in Fulton, Missouri, the museum's director, Timothy Riley, thrust a typewritten essay by Winston Churchill into his hands. Livio discovered to his amazement that in this eleven-page essay, entitled "Are We Alone in the Universe?," Churchill mused presciently about the existence of extraterrestrial life. Apparently Churchill penned the first draft of this essay, perhaps for London's *News of the World* Sunday newspaper, in 1939—when Europe was on the brink of war. He then revised it slightly in the late 1950s while staying in the south of France at the villa of his publisher, Emery Reves. Wendy Reves, the publisher's wife, passed the manuscript on to the US National Churchill Museum archives in the 1980s. Riley, who became director of the museum in May 2016, had just rediscovered the previously unpublished manuscript. The quote at the top of this chapter comes from that intriguing manuscript. Churchill sarcastically added that he did not think that "we are the highest type of mental and physical development which has ever appeared in the vast compass of space and time."

Many today think like Churchill, but speculating about it is one thing and proving that other technological civilizations exist (whether biological or AI) is another. Here we shall briefly describe a few of the searches for technosignatures that have been conducted so far, and examine ideas for what to do next.

## Electromagnetic Searches

Much of the effort to detect technosignatures to date has concentrated on searches for radio signals. The main reasons are simple: electromagnetic radiation travels at the speed of light, it is relatively easy (and cheap) to transmit and receive, and this is the type of radiation emitted by our own TV networks and radio channels. At the same time, since we don't know either the frequencies at which alien civilizations might transmit, or the nature, direction, and timing of potential signals, past searches have attempted to make some educated guesses as to where, how, and which types of signals to look for. For example, since the search for life in general has adopted a "follow the water" philosophy, the range of frequencies between 1.4 and 1.7 gigahertz (GHz), known as the "water hole," has been a favorite band of the radio quest (1.42 GHz is emitted by hydrogen atoms and 1.67 GHz by the OH component of water). Similarly, since an advanced civilization may detect the presence of life on other planets via transit spectroscopy, one idea that has been suggested was that they may also advertise their own existence (assuming they want to do so, and that they still live on the surface of a planet) while their home planet is transiting its host central star. It also goes without saying that special attention has been given to planets in the habitable zone (including the popular TRAPPIST-1 system).

Many of the radio searches to date have been conducted through various ramifications of the SETI project, whose modern incarnation was pioneered by astronomer Frank Drake, following the original, inspiring ideas by physicists Giuseppe Cocconi and Philip Morrison. The radio surveys have been using a variety of radio observatories, such as the Green Bank Telescope, the Allen Telescope Array, the Jansky Very Large Array, the Murchison Widefield Array, and the Low-Frequency Array (LOFAR). What has been found? Unfortunately, so far nothing—that is, no persistent radio transmission (at the frequencies used, down to the detection limit). However, given that the total number of methodically surveyed stars or exoplanetary systems is still very small compared to the number of stars in the galaxy, the data are grossly insufficient to enable any informed conclusions. We should note in this respect that the far side of the Moon would be an excellent site for a radio observatory, since the far side is virtually free of contamination from human radio emissions, and such an observatory would therefore permit radio searches with an unprecedented sensitivity. Radio observatories for cosmological observations have already been and are being constructed on Earth, with the goal of detecting cosmic hydrogen emission from the very early universe. It so happens that the bands of radio frequencies covered by those radio telescopes precisely overlap with the range of frequencies used for radio telecommunication on Earth. Consequently, such observatories (on the ground and even more so on the far side of the Moon) would be able (in principle) to detect the leakage of radio broadcasts from similar exo-civilizations.

Interestingly, there was one radio signal in 1977 that initially generated much excitement and was even elevated to the status of having its own name—the *Wow!* signal. The Wow! signal was recorded on August 15, 1977, by the Big Ear radio telescope of Ohio State

University, as part of the SETI project. It was discovered in the data by astronomer Jerry Ehman a few days later, and he wrote "Wow!" on the computer printout. The signal consisted of narrow-band emission close to the frequency of 1.4GHz of hydrogen, and it had an intensity that was way above the background noise. Alas, while there have been quite a few attempts to find the signal again, the latest one as recently as 2022 by the Breakthrough Listen project (see description below), all such searches have failed to detect anything. As a result, researchers agree that the Wow! signal cannot be taken as representing a genuine technosignature.

In 2022, Chinese astronomers using the Five-hundred-meter Aperture Spherical radio Telescope (FAST) detected a radio signal that, in principle, could be from an extraterrestrial civilization—that is, it was at a frequency of 1140.604 MHz, with a very narrow bandwidth. Moreover, the signal seemed to have come from the direction of a rocky exoplanet named Kepler 438b, which is in the habitable zone of a red dwarf star. Just as in previous cases, however, the signal was swiftly dismissed as being due to radio frequency interference (RFI) from Earth, with the researchers concluding: "Although we have not yet determined the exact cause of this signal, its polarization characteristic suggests that it is most likely to be attributed to RFI."

A new initiative that constitutes the most ambitious SETI venture to date was launched on July 20, 2015, and started operating in January 2016. The program, known as Breakthrough Listen, plans to survey the million closest stars to the Sun, as well as the center of the Milky Way, and even "listen" for potential messages from the one hundred galaxies that are closest to ours. The project was started with $100 million in funding from entrepreneur Yuri Milner, and with thousands of hours of observations secured on the Green Bank Telescope in the Northern Hemisphere and the Parkes telescope in the Southern Hemisphere.

In the results released through 2022, no confirmed technosignatures were found. Still, in April and May 2019, an intriguing signal at a frequency of 982.002 MHz was detected by the Parkes Radio Telescope, apparently from the direction of the star closest to the solar system, Proxima Centauri. This star is known to have at least two planets, one of which is a rocky world. The frequency was observed to drift slightly in a fashion that is often consistent with the Doppler shift of an orbiting object, although inconsistent with the motion of any known planet in the Proxima Centauri system. In an expression of the hopes of the observers, the signal was dubbed Breakthrough Listen Candidate 1 (BLC1), but some form of terrestrial radio frequency interference was still considered by veteran practitioners of SETI to have been the most likely source. By the end of 2020, follow-up observations had failed to detect the signal again, and after further analysis in 2021, the Breakthrough Listen team concluded that "BLC1 is not an extraterrestrial technosignature, but rather an electronically drifting intermodulation product of local, time-varying interferers aligned with the observing cadence."

A second type of electromagnetic signal that has been searched for was in the form of optical (and, to a lesser extent, infrared) collimated laser beams (that maintain their shapes over long distances). The idea was that an advanced civilization might prefer to use this technology for interstellar communication. In this case, however, instead of searching for long-duration signals, surveys have primarily concentrated on attempts to detect short pulses. Disappointingly, in spite of half a dozen studies having surveyed more than twenty thousand stars, no evidence for pulsed laser signals has been found.

Two of the stars that have been subjected to a SETI-type examination deserve a special mention, in spite of their rather uninspiring names: these are HD 139139 and KIC 8462852. The first one—a Sun-like star about 350 light-years away from Earth—is most likely a member of a binary stellar system, with the secondary companion being a red dwarf. HD 139139, dubbed the Random Transiter, exhibits multiple dips in its brightness (twenty-eight events over an eighty-seven-day period), somewhat similar to those caused by transiting Earth-like planets, except that the dips do not appear to be periodic. Naturally, the star became a target for technosignature searches. In particular, Breakthrough Listen observed HD 139139 with the Green Bank Telescope, but did not detect any signals.

The second star of interest, KIC 8462852, is also known as "Tabby's Star" or "Boyajian's Star" after astronomer Tabetha S. Boyajian, who was the lead author on the 2015 paper that announced the discovery of this star's irregular light fluctuations. Those fluctuations include an up to 22 percent reduction in brightness. While many explanations for the star's peculiar dimming behavior have been proposed, none of those is regarded as fully satisfactory. Theories range from the fluctuations being produced by fragments resulting from the disruption of an orphaned moon, to eclipsing megastructures constructed by an alien civilization, such as a Dyson sphere (designed so as to capture a considerable fraction of the star's power output). However, SETI searches for both optical and radio signals in the 2015–2017 time frame found no evidence for any technology-related signals from Tabby's Star. Breakthrough Listen also examined Tabby's Star in 2019 with the Automated Planet Finder at the Lick Observatory. It also produced only null results related to laser signals.

The negative results so far may seem discouraging, but we should realize that even ignoring the fact that we don't really know where and how to search, the fraction of the Milky Way that has been reached by radio-communication signals from Earth is no more than about 1 percent. To give ourselves better odds for success, we might want our signals to reach about half of all suitable exoplanets before expecting a return signal. That puts the more probable time for reception of a radio signal from another civilization in our galaxy (assuming it exists and uses this type of technology) some 1,500 years into the future. In addition, there is the distinct possibility that radio (or optical/infrared) communication might be considered archaic to an advanced life-form. The use of such technology might have been short-lived in most civilizations, and hence extremely rare over large volumes of our galaxy or the universe.

The question then is whether there is any type of generic technosignature that we could anticipate. In principle at least, energy consumption may be a hallmark of any advanced civilization, and the resultant waste heat appears to be virtually impossible to conceal. One of the most plausible long-term energy sources available to a supremely advanced technology is starlight. Powerful alien civilizations might build megastructures dubbed Dyson spheres to harvest stellar energy from one star, many stars, or even an entire galaxy. The other potential long-term energy source of an advanced species is the controlled fusion of hydrogen into heavier nuclei—a research project onto which humanity embarked in the 1940s, but where a slight net power gain (a device producing more energy than it consumed) was first confirmed only in 2024. In both cases, however, waste heat and an associated detectable mid-infrared signature would be an inevitable outcome. The point is that even with the expected higher-efficiency energy production of such an advanced civilization, the second law of thermodynamics still ensures that some processes are irreversible. One concern in employing this particular search methodology is that emission from circumstellar dust might confuse any putative signal. The hope is that natural signals would be distinguishable spectroscopically from artificial ones.

Results from the largest infrared survey to date by the Wide-field Infrared Survey Explorer (WISE) satellite were published in 2015. The researchers, led by astronomer Roger Griffith of Penn State University, examined about one hundred thousand galaxies for extreme mid-infrared (MIR) emission. They found no galaxies in their sample that could host an alien civilization that is reprocessing more than 85 percent of its starlight into the mid-infrared, and only fifty galaxies that had MIR luminosities consistent with more than 50 percent reprocessing. Perhaps most interesting, they have identified five red spiral galaxies whose combination of high MIR and low near-ultraviolet (NUV) luminosities was inconsistent with simple expectations from high rates of star formation. The NUV luminosity, dominated by young stars, typically tracks the star-formation rate, whereas the MIR luminosity, dominated by the much more abundant low-mass stars, tracks the total stellar mass. However, a more prosaic explanation for those observations, such as the presence of large amounts of internal dust, has not been ruled out. Nevertheless, such peculiar galaxies deserve follow-up observations (both by SETI, and by conventional astronomy), before we make hasty speculations about whether they represent the signature of galaxy-dominating species. These interesting findings highlight the fact that data-intensive SETI activities can and often do lead to surprising and interesting scientific discoveries that are quite unrelated to the original goal of technosignature detection.

There have been many other suggestions as to what may constitute artifact signatures indicating the potential presence of an advanced civilization. For example, the detection of various forms of atmospheric industrial pollution or short-lived radioactive products (and perhaps even a concomitant global warming). We, the authors, feel that those may not offer a high probability of detection, since those artifacts are necessarily transitory. Basically, we expect that intelligent aliens either learn how to clean up their act or they destroy themselves.

In Chapter 10, we briefly discussed the possibility of non-biological species. We noted that there is little doubt that there are chemical and metabolic limits to the size and processing power of organic brains, but that the same limits do not apply to quantum computers (or even to electronic ones). This means that the intellectual capacity and intensity of organic brains on Earth will almost certainly be eventually surpassed by some form of artificial intelligence, given that the latter is at its very early stages. The only question is when this will happen. Computer scientist Ray Kurzweil and a few other futurists think that the "singularity"—AI dominance—is no more than a few decades away. But even if it takes a few millennia, this is nothing compared to the evolutionary timescales that were needed for humans to emerge. Thousands of years are also no more than the blink of an eye compared to the lifetime of Sun-like stars, around which technological civilizations could emerge. This means that in principle at least, humans may be far inferior intellectually to alien technological civilizations.

As we have already noted, an important question that arises is related to the issue of consciousness. That is, researchers and philosophers are still debating whether consciousness is an *emergent* property—one that all sufficiently complex computational systems will eventually possess—or whether it is uniquely associated with biological brains. If machines, as "intelligent" as they may become, nevertheless lack awareness of themselves and of the world, we will probably consider them only as what philosophers refer to as "zombies," and detecting such beings, while interesting in itself, will be somewhat less exciting. On the other hand, the design of robotic systems that are effective in carrying out general-purpose real-world activities will almost certainly require the robotic brain to model not only the external world, but also its internal state, possibly leading to the spontaneous and unintended emergence of self-awareness. The question of how desire and values might emerge spontaneously in robotic/computational systems is even more mysterious and important. After all, why would an AI system explore the galaxy (or hide away) if it lacked the desire to do so?

In relation to the actual search for technosignatures, the possibility of post-human machine domination introduces yet another intriguing twist. The point we made is that organic creatures need a planetary surface (and a solvent) for the chemical reactions leading to the origin of life to take place. But if post-humans are really fully electronic intelligent beings, they will no longer need either liquid water or an atmosphere. They may even prefer a zero-gravity environment, especially for the construction of massive structures. So it may be in outer space, not on an exoplanetary surface, that non-biological minds may exist and thrive.

The type of organic human-level intelligence we are familiar with may be, generically, just a brief phase in evolution, before the machines take over. If alien intelligence has evolved along similar lines, we'd be extremely unlikely to catch it in that brief sliver of time when it was still embodied in the organic form. Particularly, were we to detect an extraterrestrial technological civilization, it would be far more likely to be electronic, where the dominant creatures aren't flesh and blood.

So whereas the focus of the search for extraterrestrial intelligence has been so far on radio or optical signals, we should be more alert to evidence for non-natural engineering projects, such as "Dyson spheres," constructed so as to harvest a large fraction of stellar power, and even to the potential presence of alien artifacts keeping out of sight within our own solar system. Investigating this last possibility is, in fact, the goal of the *Galileo Project*, headed by astronomer Avi Loeb of Harvard University, who is a passionate advocate of the search for technosignatures. In July 2023, Loeb and his team discovered remains from a meteor that landed in the waters off of Papua New Guinea in 2014, in the form of tiny metal spherules. Based on the meteor's recorded speed, Loeb and his collaborators concluded that the meteor was likely of interstellar origin. The spherules were 0.05–1.3 millimeters in size, about 850 of them in total. From an analysis of the composition of about fifty spherules, Loeb and his team claimed that five of them (which showed a high percentage of beryllium, lanthanum, and uranium) originated outside the solar system. Initially, Loeb speculated that these might be fragments of a spacecraft from another civilization or from some other technological gadget. After the spherules were taken to Harvard for further analysis, the team stated in a publication that the unusual composition "could have originated from a highly differentiated magma ocean of a planet with an iron core outside the solar system or from more exotic sources." A subsequent independent study by Patricio Gallardo of the University of Chicago suggested that the spherules' composition is consistent with coal ash contaminants—that is, human-generated industrial pollution. Though Loeb's team disputed this claim, most astronomers are extremely skeptical about the suggestion that the spherules are associated with an alien spacecraft. Many even regarded Loeb's pronouncements as being so outlandish that they raised the concern that such speculative statements would create a false impression of how solid science is truly carried out. We should nevertheless note that we, the authors, support the general idea of searching for potential alien artifacts within the solar system. Unlike electromagnetic signals, which could be encrypted to such a degree that we would not even recognize them as being artificially created, technological artifacts (if they are found) might be easier to identify as such. Even if ET artifacts are not found, such searches might lead, just like searches for electromagnetic ET signals, to the discovery of scientifically interesting surprises unrelated to the original goal.

Another astronomical body that has generated considerable discussion and controversy was an interstellar object detected on October 19, 2017, zipping through the solar system. It is best known as Oumuamua, which roughly translates from Hawaiian to "first distant messenger," or "scout." In addition to having clearly originated from outside the solar system on the basis of its speed and trajectory, what made Oumuamua intriguing was its unusual cigar or pancake shape (being 300 to 3,000 feet long, and only 115 to 550 feet wide and thick). It also showed no sign of a comet-like coma—the nebulous envelope formed by sublimation from a comet's nucleus. For comparison, comets are typically a few miles across. As a result of its peculiar characteristics, the object was initially thought to be an asteroid, but further analysis in 2018 showed that it was also exhibiting a non-gravitational acceleration as it was heading out, away from the solar system. Given Oumuamua's unique properties, it wasn't surprising perhaps that Loeb suggested that the rock could be an alien probe produced by an extraterrestrial technology. However, radio observations by the SETI Institute's Allen Telescope Array and by Breakthrough Listen's Green Bank Telescope detected no unusual radio signals. Moreover, in a paper published in March 2023, astrochemist Jennifer Bergner of the University of California, Berkeley, and astronomer Darryl Seligman of Cornell University suggested that the object is consistent with being a tiny comet, accelerated by very small amounts of hydrogen gas gushing from an icy core. This is not to say that everybody agrees that this is the correct model, but most astronomers do think that despite its weirdness, it doesn't hold that this is an alien spaceship. Basically, it doesn't pass the test of Sagan's maxim: "Extraordinary claims require extraordinary evidence." By the way, a second interstellar object was discovered in 2019 by amateur astronomer Gennadiy Borisov (and named "2I/Borisov"), but in this case the object was clearly a rogue comet.

Overall, we are left with no convincing detection to date of any technosignature, but we also realize that perhaps our approach has been somewhat misconceived. Together with a few other researchers, such as University of Cambridge astrophysicist Martin Rees, we, the authors, now think that if SETI were to succeed, it would be unlikely that the signal it observes would be a simple, decodable message. Rather, it may be the unintended by-product (or perhaps even the result of an accident or malfunction) of some super-complex machine, far beyond our comprehension. Even if alien messages were transmitted, we may not recognize them as artificial because we may not know how to decipher them, in the same way that a veteran radio engineer familiar only with amplitude modulation (AM) might have a hard time decoding modern wireless, digitally encoded communication. Indeed, data compression techniques today aim to make signals as similar to irregular noise as possible.

To conclude: conjectures about advanced, intelligent technological life involve many more uncertainties than those about the nature of simple life. In particular, SETI searches so far may have been misguided. It is quite possible that alien technological civilizations will not be organic or biological. As such, they may not remain on the surface of the exoplanet on which their biological precursors lived and evolved, and we will certainly not be able to fathom or predict their motives, intentions, or behavior. As a result, attempts to estimate the most uncertain factors in the Drake equation may turn out not to be useful in the searches for technosignatures.

We cannot end this chapter without at least mentioning claims of sightings of Unidentified Flying Objects (UFOs), or, as they are now referred to, Unidentified Aerial Phenomena (UAPs). The topic cannot be ignored, especially since on July 26, 2023, three former military officials told a US congressional House oversight committee that they believe the US government knows much more about UAPs than it is telling the public. They gave baffling testimony about what they regarded as unexplained object sightings and even about governmental possession of what they called "nonhuman" biological matter, though they did not provide any evidence to support those allegations. In fact, as many experts have pointed out over the years, many such phenomena might be attributable to various types of balloons, drones, atmospheric events, optical illusions, blinking lights of commercial airliners, or just pure hoaxes. Indeed, the Pentagon's reaction was that they have seen no evidence linking UAPs to alien activity, though they are not a priori ruling out such possibilities. An independent group of sixteen experts convened by NASA also found no evidence that UAPs are extraterrestrial in nature, even though some events have defied explanation. We, the authors, have not seen in all the published claims anything that rises to the level of extraordinary evidence for the existence of alien technological civilizations. To us, therefore, stories of UFOs or UAPs remain, for the moment, an interesting cultural artifact, rather than representing scientific discoveries. In this respect, the advice from Jason Wright, an astrobiologist at Penn State University, seems appropriate: "Stay skeptical, but not cynical."

This brings us to a brief summary (in the next chapter) of where we think we are now in the quest for producing a living cell from chemistry in the laboratory, and in the search for extraterrestrial life.

# CHAPTER 13

# Epilogue

## An Imminent Breakthrough?

*Therefore the problem is not so much that of seeing what no one has yet seen, but rather of thinking in the case of something seen by everyone that which no one has yet thought.*

—Arthur Schopenhauer, *Parerga und Paralipomena*

How did life on Earth begin? Has life emerged elsewhere in our galaxy? These were the two major, fundamental questions with which we started this book. The answer feels close, but we are still unable to give a definitive reply to either of these questions. In fact, we should realize that we may never know what *precisely* happened on Earth some four billion years ago, when prebiotic chemistry formed the very first protocells, or when those primitive cells started evolving into something like modern life. Nevertheless, we have seen how using knowledge acquired through ingenious chemical experimentation, geological studies, advanced astronomical observations, and imaginative theorizing, researchers have managed to delineate a very plausible pathway (admittedly incomplete) leading from the formation of Earth to the appearance of early cells. At the same time, we have also presented the astounding findings of astronomers and astrobiologists in the last three decades—discoveries that have brought us to the verge of being able to detect extraterrestrial life (if it is not exceedingly rare), or at least to the ability to place meaningful statistical constraints on how rare such life may be.

Searching for the origin of life, or for extraterrestrial life, may seem esoteric and far removed from the problems we are facing in our everyday lives, but such has always been the nature of basic research. The renowned Swedish physical chemist Svante Arrhenius published in 1909 a fascinating book entitled *The Life of the Universe.* He ended that book with the following thoughtful comments:

> *Nothing can be more mistaken than to state that time spent in theorizing on cosmogonical problems is wasted, and that we shall never advance beyond the knowledge of the ancient philosophers. . . . Culture and civilization expand, when the human race advances. And we find in particular that the scientist has, in all ages, spoken for humanity.*

Judging from the history of science, we agree with the statement that there are two types of science: there is applied science, and *not yet* applied science. A famous quote in this respect is attributed by the Irish historian W. E. H. Lecky to the great nineteenth-century experimentalist in electromagnetism Michael Faraday. It concerns an exchange between Faraday and William Gladstone, who was at the time the British Chancellor of the Exchequer, responsible for the national budget. Lecky writes in his 1899 book *Democracy and Liberty*: "An intimate friend of Faraday once described to me how, when Faraday was endeavoring to explain to Gladstone and several others an important new discovery in science Gladstone's only commentary was 'but, after all, what use is it?' 'Why, sir,' replied Faraday, 'there is every probability that you will soon be able to tax it!'" Indeed, when James Watson and Francis Crick started constructing models of the structure of DNA, they could not have imagined that their work would give birth, for example, to a DNA biotechnology industry that has revolutionized medicine and is worth hundreds of billions of dollars.

So, where are we right now in the attempts to synthesize a living cell in the laboratory, and in the observational endeavors to detect extraterrestrial life? In both areas, there have been huge advances in the past decade, and there is a growing feeling of excitement that many of the remaining questions will also soon be answered. Perhaps the most wonderful aspect of this progress is that new questions have arisen, which until recently we did not even know enough to ask. Here is a very brief summary of the current state of the science of the origin and prevalence of life, together with a few of the questions we have only recently learned enough to ask.

## Origin of Life

The most crucial unsolved problems on the road from pure chemistry to biology can be divided into three groups: the remaining issues in the synthesis of the building blocks of biology, the obstacles associated with the assembly of the first cells, and the questions regarding the subsequent evolution of early life. In terms of nucleotide synthesis, we now have in hand a concise and efficient chemical pathway to the production of the pyrimidine nucleotides (the basic building blocks C and U). As we have seen, we can trace this pathway from the capture of cyanide from the atmosphere as ferrocyanide salts that accumulate with time to generate a reservoir of starting material. Later thermal processing of this material by the heat and pressure of lava flows or meteorite impacts, followed by cooling and later groundwater leaching, would generate a highly concentrated solution of reactive feedstocks such as cyanide and cyanamide. Once brought to the surface and exposed to UV light, together with the help of sulfite from volcanic outgassing, simple sugars would be formed, followed by the synthesis of the intermediate 2AO and then the remarkable crystalline intermediate RAO. Reaction with cyanoacetylene, itself derived from a crystalline reservoir, generates the anhydro-C nucleoside, which reacts with sulfide to form the α-anomer of thio-C (in which one of the oxygen atoms has been replaced with a sulfur atom). Exposure to UV light then flips the orientation of the nucleobase to give the biologically relevant anomer β-2-thio-C, which can then deaminate (removing the amino group) to give rise to 2-thio-U. These sulfur-containing versions of C and U can then lose the sulfur to generate the canonical pyrimidine nucleosides. But here is an example of a question that we didn't even know enough to ask before this pathway had been worked out. Was the primordial genetic alphabet based on the sulfur-containing pyrimidines, or on their non-sulfur-containing "modern" versions? The advantages and disadvantages of an earlier version of RNA based on the sulfur-containing nucleotides is now the subject of intense experimental work in several laboratories. As we have shown, a very satisfying aspect of the above chemistry, which is based on cyanide, sulfur, and UV light, is that it also explains the synthesis of at least eight of the amino acids that are universally used as building blocks for protein synthesis in modern biology. Having the amino acids and the nucleotides emerge from a common core chemistry is a surprising and truly delightful outcome of these chemical investigations.

This brief summary of the elegant path to the C and U pyrimidines naturally begs the question of the purine nucleosides (A and G), the synthesis of which is the biggest remaining gap in our knowledge of the origins of RNA. Needless to say, this is also the subject of intensive experimental study and debate, with most researchers sharing the hope that this missing link will soon be found. The second major area of our ignorance is the chemistry of phosphate activation (how to attach a phosphate to the nucleosides). We already have a variety of chemical means of activating phosphates that are not prebiotically relevant, and we have a few (such as one known as isocyanide chemistry) that have at least some tenuous prebiotic plausibility. Nevertheless it is widely believed that the discovery of a robust and realistic chemistry that could drive the synthesis of activated nucleotides would fill a major gap in our understanding of RNA synthesis, copying, and replication. Finally, the synthesis of the lipid components of protocell membranes remains largely mysterious and is clearly an aspect of prebiotic chemistry that deserves more attention. In conclusion, we're not there yet, but we are approaching a stage where we will understand chemically efficient pathways to all of the basic building blocks of biology.

Other important questions still need to be answered. In particular, the chemistry we have described needs to be placed into the context of the geology of the early Earth. At this stage we only have bits and pieces of likely scenarios. One gratifying example is the well-modeled plausibility of the accumulation of ferrocyanide salts in alkaline carbonate lakes, together with the free phosphate required for making nucleotides. There is also the beautiful, recently proposed idea that the mirror-image forms of the nucleotide precursor RAO could be separated by crystallization on a surface composed of magnetite. Magnetite is a common mineral that forms in lakes containing dissolved iron that oxidizes due to exposure to UV light, and then precipitates as iron complexes that transform into magnetite. In the presence of a magnetic field, such as Earth's natural magnetic field, the microscopic grains of magnetite become oriented by the magnetic field. The recent experimental demonstration that one of the two mirror-image forms of RAO will specifically crystallize on such a magnetized surface is an exciting discovery that offers a potential solution to the problem of attaining homochiral nucleotides in an abiotic environment. Experiments are currently being carried out to rigorously test the plausibility of this process in a realistic early-Earth environment.

A final and very difficult remaining problem concerns the probability of having all of these separate geochemical scenarios linked together in the right sequence so as to generate all the required building blocks of life together, at the same time, in the same place. For example, accumulated reservoirs of ferrocyanide salts might form commonly, but most of the time they also might be washed away or destroyed, and only rarely be processed correctly to give the compounds needed for subsequent steps. In addition, those compounds have to be delivered to an environment where those next steps could occur. Similarly, reservoirs of pure crystalline RAO and CV-DCI might build up in many locations, but we don't know how often they would subsequently be dissolved and combined together, under the right conditions, to generate the next precursor of the nucleotides. The problem of how to correctly model such a series of accumulations, transport processes, and chemical reactions remains a huge challenge for the field.

Then there is the question of how close we are to understanding how the first protocells assembled. Let's assume that an environment with all the necessary components existed somewhere on the early Earth. What then? Given the presence of activated nucleotides, the spontaneous assembly of RNA oligonucleotides (short nucleic acid polymers) seems straightforward, since several different kinds of changing environments (wet-dry cycles, freeze-thaw cycles) can lead to polymerization. Similarly, given the presence of a high-enough concentration of lipids, which could be as simple as fatty acids, the self-assembly of membrane vesicles is virtually inevitable. As long as these processes can happen together, the assembly of the basic protocell structure, that is, RNA encapsulated within membrane vesicles, is not difficult to envisage. Even the trickier aspects of protocell reproduction seem close to being understood, although to be sure there are still gaps in our knowledge. As we have seen, multiple distinct processes can lead to the growth and subsequent division of simple membrane vesicles under conditions that could easily occur on the early Earth. More work is required to achieve a similar understanding of RNA replication, but basic copying chemistry is becoming well understood, and ideas about how to use this chemistry to drive replication are being intensively studied. A big question here is whether replication, like vesicle growth and division, must be driven by environmental fluctuations, such as hot-cold cycles, or if it could proceed under gentle and almost constant conditions. Another key open question is how RNA replication could proceed with sufficient fidelity to allow for the transmission of a useful amount of genetic information, that is, the information required to encode a beneficial RNA catalyst. Finally, perhaps the thorniest unresolved problem of all is how RNA replication could occur under conditions compatible with vesicle growth and division. After all, everything has to work together for protocells to reproduce and begin to evolve. Whether the solution to this puzzle lies in finding a more robust membrane composition, or in discovering a different way to catalyze RNA replication, or some combination of the two, is still unknown, but we suspect that an answer will emerge in the not-too-distant future.

Once a population of reproducing protocells has become established in some favorable local environment, the question is what we might expect to have happened next. Instant extinction might be the most likely outcome, given the early Earth hazards of volcanic eruptions and meteorite impacts. It may be that protocell populations had to emerge many times in different locations, before one was lucky enough to survive sufficiently long to evolve into a more robust form of life that could disperse around the planet, establishing footholds so widespread that no single catastrophic event could destroy this incipient form of primordial life.

What would the earliest steps of Darwinian evolution have looked like? It is widely thought that strong selection for more efficient protocell reproduction would be focused on RNA replication. The evolution of a ribozyme that could catalyze RNA copying and even replication, commonly referred to as an RNA replicase, is certainly one possibility, and it makes sense that a catalyst that improved on RNA copying chemistry would help to make RNA replication more robust. But improved RNA replication would only be selected for if it upgraded the reproduction or survival of the protocell itself. This could happen if RNA replication was coupled to membrane growth, for example through osmotic pressure. However, another possibility is that the first ribozyme did something else entirely. Once there is some first ribozyme that has an advantageous effect on the protocell, then replicating that RNA sequence in a more efficient and more accurate way would be strongly selected for. The evolution of a more efficient replication machinery would then enable a primordial cell to maintain a larger genome that could code for additional ribozymes carrying out yet other new functions. Such a cascade effect could, step by step, enable an expanding population of primordial cells to explore different environments. How would these early cells have spread between distant environments? One popular idea is that these cells could have been caught up in the aerosol droplets that result from breaking waves, and then carried for long distances by the wind. Another hypothesis is that primitive cells could have dried up and been blown about as dust particles. Whenever one landed in a nutrient-filled pond or lake, it would have rehydrated and recommenced its cycle of growth and division. Although just hypotheses at present, these ideas can all be tested in laboratory experiments, which should give us a better idea of how life initially spread across and colonized the entire early planet Earth.

At this point we can start thinking about how, once the process of Darwinian evolution had become well established, the complex systems that are characteristic of all modern forms of biology arose. These systems include cellular metabolism, the translation system that allows for the synthesis of coded protein enzymes, the amazingly complicated proteins that mediate all transport across cell membranes, and of course the specialization of information storage in DNA. Indeed, how the relatively simple cells of the RNA World gave rise to the much more complicated cells of today's biology is largely a mystery, or rather a set of mysteries. Even though we lack direct evidence, there are a few simple logical considerations that place constraints on how modern life developed, and that may help to guide future research. Let's begin by thinking about the basic protocell structure: replicating RNA fragments encapsulated within a membrane vesicle. This membrane had to be leaky enough to allow for nutrients such as nucleotides, made in the external environment, to enter the cell spontaneously. This implies that any useful nutrient made inside the cells would be just as likely to leak out as to be used internally, and therefore there would be no use in having an internal metabolism until the membrane itself became less leaky. Why would that happen? As we have seen, the synthesis of even a small amount of phospholipid, perhaps by a ribozyme with the correct catalytic activity, would lead to enhanced membrane growth. This would result in an evolutionary arms race as cells competed to make more phospholipids, which in turn would have led to a changing membrane composition with important consequences for the evolving protocells. As the membrane became progressively more enriched with phospholipids, it would have become less permeable, making it harder for the cell to import nutrients from the environment, but by the same token making it worthwhile to synthesize nutrients internally, that is, by its own cellular metabolism. Modern cellular metabolism is a highly complex network of hundreds to thousands of enzyme-catalyzed reactions. To understand the evolution of this network we must remember that every novel innovation—every new RNA enzyme that catalyzes a metabolic reaction—had to result in a selective advantage for the host cell. The earliest metabolic innovations presumably increased the synthesis of the activated nucleotides required for RNA replication, and/or the synthesis of membrane components. But how cells could gradually transition from a complete reliance on environmentally supplied nutrients (a heterotrophic lifestyle) to being able to make all of their own components from simple, abundant, and readily available starting materials is not easy to understand. This is one of the most interesting current questions about the evolution of early life.

Another intriguing issue concerns the replacement of a rudimentary set of metabolic reactions, all catalyzed by RNA enzymes, by protein enzymes, and the parallel emergence of the protein machinery that controls transport across the cell membrane. Most theories of the origin of protein synthesis have focused on the origin of the genetic code. This process remains quite obscure, although there are certain patterns in the code that point to a stepwise emergence of the full coding system. It seems likely that a subset of the canonical twenty amino acids were used to code for the first proteins, with additional amino acids being added to the code at later times. The ribosome itself is an incredibly complicated molecular apparatus, with two main subunits, one of which catalyzes the formation of new peptide bonds, while the other directs the actual coding. The peptide-synthesizing activity may have come first, if non-coded peptides conferred some benefit, with mRNA-directed coding being added to the system at a later time. The substrates that the ribosome uses for peptide synthesis are also intricate: they are particular RNA molecules that have specific amino acids attached at one end. The ribosome could not have evolved unless its substrates already existed, but why would primitive cells have been making aminoacylated (amino acid attached to RNA) RNAs? One possibility is that aminoacylated RNAs played a prior role in facilitating the assembly of ribozymes, before they were co-opted for their role in peptide synthesis. According to this hypothesis the first enzymes were ribozymes made only of RNA; improved chimeric RNA–amino acid enzymes then arose, and finally these gave way to all-peptide (protein) enzymes. The ability to generate coded proteins with hydrophobic surfaces also allowed for the evolution of proteins that could sit within the membrane, where they could act as channels, pores, and pumps to facilitate the import and export of molecules as needed by the cell.

Finally, there is DNA itself. The advantage of DNA as a medium for storing large amounts of information is clear-cut, because DNA is so much more stable than RNA with regard to chemical degradation. In addition, the specialization of the function of DNA for information storage and genome replication, and RNA for its functional roles, would seem to be inherently advantageous, since each molecule could perform its own functions in an optimal manner, avoiding the compromises associated with one biopolymer trying to carry out multiple functions. An important question is whether DNA evolved to be the primary genetic storage material early or late in evolution—also, whether the switch to DNA for information storage occurred quickly, or through a gradual transition. One argument in favor of at least the possibility of an early role for DNA is that the synthesis of deoxynucleotides may have occurred together with the synthesis of ribonucleotides. If ribo- and deoxyribo-nucleotides were synthesized together, it is possible that primordial cells contained a mixed RNA/DNA polymer. The specificity required for making "pure" RNA may have had to wait for the evolution of a ribozyme RNA polymerase (an RNA enzyme that can catalyze RNA synthesis). In that case, a mutant enzyme with a preference for DNA synthesis might have enabled a relatively rapid switch to the synthesis of DNA for information storage, while RNA retained its ancestral function for catalysis. While we may never know the exact sequence of events that took place so long ago during the early evolution of life, the space of possibilities can be demonstrated through laboratory evolution experiments. Being able to synthesize simple cells at various levels of complexity is certainly a very exciting prospect for the future, and it may allow us to trace out a plausible pathway for the evolutionary transition from the first, simplest protocells, to cells with the complexity of modern bacteria.

## Extraterrestrial Life

On the astronomy front things are, perhaps, even closer to a major breakthrough. As we have described in this book, we have already witnessed in the past few years multiple (admittedly premature) claims for potential detections of extraterrestrial biosignatures—one in the solar system itself (in the form of *phosphine* on Venus), and one in the guise of the putative detection of *dimethyl sulfide* on the hycean extrasolar planet K2-18b. The important point is that both of these claims will soon be tested by planned observations. Moreover, there have even been assertions (albeit extremely speculative) that technosignatures may have been discovered, one in the form of meteorite relics on Earth, supposedly originating from an alien spacecraft, and the other through the detection of a puzzling interstellar visitor object (called Oumuamua) to the solar system. While most astronomers are convinced that none of these particular "detections" truly represents either a bona fide biosignature or a technosignature, astronomers do appear to be already well positioned to be able to detect extraterrestrial life (assuming it exists) within a mere decade or two. This expectation becomes especially likely given the emphasis put by the latest Decadal Survey on achieving precisely such a detection. For example, building upon studies conducted for two earlier mission concepts called the Large Ultraviolet Optical Infrared Surveyor (LUVOIR) and Habitable Exoplanet Observatory (HabEx), NASA is planning to launch (perhaps around 2040) the *Habitable Worlds Observatory* (*HWO*), which will search in infrared, optical, and ultraviolet light for the biosignatures (and perhaps even technosignatures) of alien life on about twenty-five potentially habitable planets around stars like our Sun. If successful, this observatory will be nothing short of a technological marvel. To be able to image an extrasolar Earth-like planet—a tiny object tens of light-years away—HWO will have to be extraordinarily steady, that is, it would have to maintain a phenomenal stability.

It is difficult to predict what the response will be to a discovery of extraterrestrial life, let alone to the detection of a technosignature. Recall, for instance, the reaction to the (in retrospect false) announcement that life had been discovered in the ALH84001 meteorite from

Mars. The *New York Times* ran on its front page of August 7, 1996, the headline: "Clues in Meteorite Seem to Show Signs of Life on Mars Long Ago." Still, since that particular news item concerned only the potential detection of primitive life (and even that, on a solar system object), the excitement was felt primarily within the scientific community, with the general public showing no more than a modest level of interest. Undoubtedly a sign of an intelligent civilization is likely to generate a much stronger response, but the intensity and nature of that reaction may depend on other factors, such as the perceived distance between Earth and that advanced civilization. The excitement may also be accompanied by nervousness and fear. After all, even the famed astrophysicist Stephen Hawking advised that humanity should be wary of seeking out contact with any alien civilizations. In an online video, he said, "Meeting an advanced civilization could be like Native Americans encountering Columbus," admonishingly adding, "That didn't turn out so well." Religious beliefs are also likely to be affected, although perhaps not to the extent you might imagine. For example, the director of the Vatican Observatory, Brother Guy J. Consolmagno, SJ, expressed his personal optimism about the resilience of religions by noting: "If your religion has survived millennia—if it can handle Copernicus, Galileo, and even Darwin—then E.T. should eventually prove palatable."

Could it be, in spite of the fact that there are as many as hundreds of millions of habitable exoplanets in our galaxy alone, and the number of galaxies in the observable universe is in the trillions, that Earth is the only place with intelligent life? With any form of life? The problem with attempts to answer these questions is that we have no idea about either the probability of the spontaneous emergence of life, or the probability of primitive life evolving to intelligent life. The experimental and observational evidence we presented in Chapters 2–6 and our discussion earlier in this chapter demonstrate that many of the building blocks of biology can be produced by the chemistry we expect to have been realized under conditions similar to those of the early Earth. However, we have also shown that the entire sequence of required steps necessitates the buildup of particular chemical reservoirs that must become available at specific points in time, at the appropriate locations. These prerequisites make it (at least currently) virtually impossible to estimate the probability for the occurrence of the entire process. Consequently, the answer to the "Are we alone?" question will have to come from astronomy, and finding that answer may not be easy, unless we get really lucky, or if cosmic life is really ubiquitous. The crucial point is that, as physicist Philip Morrison once said: "The probability of success is difficult to estimate; but if we never search the chance of success is zero."

# ACKNOWLEDGMENTS

It would be virtually impossible to name all the collaborators, colleagues, and authors who have contributed, directly or indirectly, to the writing of this book. The list below should therefore be regarded only as partial and representative rather than complete. We are especially grateful to Fred Adams, Philip Armitage, John Barrow, Anat Bashan, Sagi Ben-Ami, David Catling, Irene Chen, Adam Frank, Patrick Godon, Andrew King, Ram Krishnamurthy, Doron Lancet, Stephen Lepp, Jack Lissauer, Avi Loeb, Stephen Lubow, Renu Malhotra, Sheref Mansy, Rebecca Martin, Michel Mayor, Peter McCullough, Eran Ofek, Jim Pringle, Fred Rasio, Martin Rees, Dimitar Sasselov, Hilke Schlichting, Sara Seager, Seth Shostak, Lionel Siess, Joe Silk, Jeremy Smallwood, Noam Soker, Massimo Stiavelli, John Sutherland, Jill Tarter, Chris Tout, Jeff Valenti, Eva Villaver, Ada Yonath, and Lijun Zhou.

Special thanks are due to our editor, T. J. Kelleher, and to Kristen Kim, Lara Heimert, and the entire production team at Basic Books.

# SELECTED FURTHER READING

**Chapter 1. A Freak Chemical Accident or a Cosmic Imperative?**
*Annotated Bibliography*

A technical, comprehensive account of many aspects of the emergence and characteristics of life on Earth and of the search for life in the universe can be found in Lingam and Loeb (2021). A more popular description of the origins and evolution of life on Earth is Ward and Kirschvink (2016). The topic of the origin of life is also discussed in a nontechnical fashion in Deamer (2020). Joyce and Szostak (2018) present a discussion on the two key components of protocells: a self-replicating nucleic acid genome and a self-replicating membrane. A simple introduction to astrobiology as a discipline is Plaxco and Gross (2021). A more personal account of the search for extraterrestrial life by a leading researcher is Seager (2020). An interesting, diverse collection of essays related to the search for extraterrestrial life is Al-Khalili (2016). Erwin Schrödinger's short book *What Is Life?*, which inspired the initiation of much of modern research, has been reprinted many times, for example, in Schrödinger (2018). A more recent view on the general question of the nature and characteristics of life is presented in Nurse (2020). The idea that life may be a cosmic imperative was expressed in de Duve (2011). While England (2013) made the speculative suggestion that he had uncovered even the physics that is driving the origin and evolution of life, many are not convinced. The historical debate on the question of the "plurality of inhabited worlds" is comprehensively discussed by Dick (1980). Greene's (2020) beautiful book on the exploration of the cosmos includes a brief discussion on the origin of life. An insightful perspective on the laws that govern the universe, and how those laws allowed for the emergence of life, is presented in Rees (2000). A statistical analysis examining whether the early emergence of life on Earth means that life is common was presented in Spiegel and Turner (2012). An interesting collection of articles discussing whether the physical constants in our universe are somehow fine-tuned to allow for complexity and life to emerge can be found in Sloan et al. (2020). Tyson and Trefil (2021) present a very engaging tour of a variety of cosmic questions, including that of cosmic life. Green (2023) gives a charming fusion of science and science fiction in her description of the search for cosmic life.

*References*

Al-Khalili, J., editor, 2016, *Aliens: The World's Leading Scientists on the Search for Extraterrestrial Life* (New York: Picador).

Deamer, D. W., 2020, *Origin of Life: What Everyone Needs to Know* (Oxford: Oxford University Press).

Dick, S. J., 1980, "The Origins of the Extraterrestrial Life Debate and Its Relation to the Scientific Revolution," *Journal of the History of Ideas* 41, no. 1 (January–March).

de Duve, C., 2011, "Life as a Cosmic Imperative?," *Philosophical Transactions of the Royal Society A* 369, no. 1936 (February).

England, J. L., 2013, "Statistical Physics of Self-Replication," *Journal of Chemical Physics* 139, no. 12 (September).

Green, J., 2023, *The Possibility of Life: Science, Imagination, and Our Quest for Kinship in the Cosmos* (New York: Hanover Square Press).

Greene, B., 2020, *Until the End of Time: Mind, Matter, and Our Search for Meaning in an Evolving Universe* (New York: Alfred A. Knopf).

Joyce, G. F., and Szostak, J. W., 2018, "Protocells and RNA Self-Replication," *Cold Spring Harbor Perspectives in Biology* 10, no. 9 (September).

Lingam, M., and Loeb, A., 2021, *Life in the Cosmos: From Biosignatures to Technosignatures* (Cambridge, MA: Harvard University Press).

Nurse, P., 2020, *What Is Life? Understand Biology in Five Steps* (Oxford: David Fickling Books).

Plaxco, K. W., and Gross, M., 2021, *Astrobiology: An Introduction* (Baltimore: Johns Hopkins University Press).

Rees, M., 2000, *Just Six Numbers: The Deep Forces That Shape the Universe* (New York: Basic Books).

Schrödinger, E., 2018, *What Is Life? With Mind and Matter and Autobiographical Sketches* (Cambridge: Cambridge University Press).

Seager, S., 2020, *The Smallest Lights in the Universe: A Memoir* (New York: Crown).

Sloan, D., Batista, R. A., Hicks, M. T., and Davies, R., 2020, *Fine-Tuning in the Physical Universe* (Cambridge: Cambridge University Press).

Spiegel, D. S., and Turner, E. L., 2012, "Bayesian Analysis of the Astrobiological Implications of Life's Early Emergence on Earth," *Proceedings of the National Academy of Sciences USA* 109, no. 2 (January).

Tyson, N. D., and Trefil, J., 2021, *Cosmic Queries: StarTalk's Guide to Who We Are, How We Got Here, and Where We're Going* (Washington, DC: National Geographic).

Ward, P., and Kirschvink, J., 2016, *A New History of Life: The Radical New Discoveries About the Origins and Evolution of Life on Earth* (New York: Bloomsbury Publishing).

## Chapter 2. The Origin of Life: The RNA World

*Annotated Bibliography*

A concise, popular summary of some of the history, ideas, and studies concerning the origin of life on Earth was presented by Marshall (2021). Brief professional reviews of research conducted in the past few years on the topic of the origin of life on Earth are presented by Sutherland (2016, 2017) and by Szostak (2017a, 2017b) and Joyce and Szostak (2018). The crucial role of "systems chemistry" in the emergence of life on the early Earth was briefly explained in Szostak (2009). The fact that RNA seems to always win against potential alternative molecules was demonstrated experimentally in Kim et al. (2021). A potential model for primordial RNA replication was suggested by Zhou, Ding, and Szostak (2021). As we have noted, despite what we regard as compelling evidence, not all researchers agree with the scenario we have presented for the origin of life.

*References*

Joyce, G. F., and Szostak, J. W., 2018, "Protocells and RNA Self-Replication," *Cold Spring Harbor Perspectives in Biology* 10, no. 9 (September).

Kim, S. C., et al., 2021, "The Emergence of RNA from the Heterogeneous Products of Prebiotic Nucleotide Synthesis," *Journal of the American Chemical Society* 143, no. 9 (March).

Marshall, M., 2021, "BBC Earth: The Secret of How Life on Earth Began," BBC, October 31, 2016, reposted on personal webpage, https://www.michaelcmarshall.com/blog/bbc-earth-the-secret-of-how-life-on-earth-began.

Sutherland, J. D., 2016, "The Origin of Life—Out of the Blue," *Angewandte Chemie International Edition* 55, no. 1 (January).

Sutherland, J. D., 2017, "Studies on the Origin of Life—the End of the Beginning," *Nature Reviews Chemistry* 1, no. 12.

Szostak, J. W., 2009, "Systems Chemistry on Early Earth," *Nature* 459 (May 14).

Szostak, J. W., 2017a, "The Narrow Road to the Deep Past: In Search of the Chemistry of the Origin of Life," *Angewandte Chemie International Edition* 56, no. 37 (September).

Szostak, J. W., 2017b, "The Origin of Life on Earth and the Design of Alternative Life Forms," *Molecular Frontiers Journal* 1, no. 2 (December).

Zhou, L., Ding, D., and Szostak, J. W., 2021, "The Virtual Circular Genome Model for Primordial RNA Replication," *RNA* 27, no. 1 (January).

## Chapter 3. The Origin of Life: From Chemistry to Biology

*Annotated Bibliography*

The potential path from chemistry to biology in the origin of life on Earth was reviewed by Szostak (2017). Gollihar, Levy, and Ellington (2014) explained that there may be many paths leading to the first self-replicating system. The crucial role played by ultraviolet light in both the photochemical synthesis of prebiotic molecules and in the selectivity for molecules that successfully function in biology is described by Green, Xu, and Sutherland (2021), and the common origin of the precursors of the molecules of life is discussed in Patel et al. (2015).

*References*

Gollihar, J., Levy, M., and Ellington, A., 2014, "Many Paths to the Origin of Life," *Science* 343 (January 17).

Green, N. J., Xu, J., and Sutherland, J. D., 2021, "Illuminating Life's Origins: UV Photochemistry in Abiotic Synthesis of Biomolecules," *Journal of the American Chemical Society* 143, no. 19 (May).

Patel, B. H., Percivalle, C., Ritson, D. J., Duffy, C. D., and Sutherland, J. D., 2015, "Common Origins of RNA, Protein and Lipid Precursors in a Cyanosulfidic Protometabolism," *Nature Chemistry* 7, no. 4 (April).

Sutherland, J. D., 2016, "The Origin of Life—Out of the Blue," *Angewandte Chemie International Edition* 55, no. 1 (January).

Szostak, J. W., 2017, "The Narrow Road to the Deep Past: In Search of the Chemistry of the Origin of Life," *Angewandte Chemie International Edition* 56, no. 37 (September).

**Chapter 4. The Origin of Life: Amino Acids and Peptides**

*Annotated Bibliography*

For a general introduction to amino acids, see, for example, Nelson and Cox (2021). The series of chemical reactions that produce an amino acid from an aldehyde was discovered by Strecker (1850). Professional works on peptide ligation are Canavelli et al. (2019), and Foden et al. (2020). The simultaneous synthesis of the precursors of ribonucleotides and amino acids is discussed in Ritson and Sutherland (2013).

*References*

Canavelli P., Islam S., and Powner, M. W., 2019, "Peptide Ligation by Chemoselective Aminonitrile Coupling in Water," *Nature* 571 (July 25).

Foden, C. S., Islam, S., Fernández-García, C., Maugeri, L., Sheppard, T. D., and Powner, M. W., 2020, "Prebiotic Synthesis of Cysteine Peptides That Catalyze Peptide Ligation in Neutral Water," *Science* 370 (November 13).

Nelson, D. L., and Cox, M. M., 2021, *Lehninger Principles of Biochemistry*, 8th ed. (New York: W. H. Freeman).

Ritson, D. J., and Sutherland, J. D., 2013, "Synthesis of Aldehydic Ribonucleotide and Amino Acid Precursors by Photoredox Chemistry," *Angewandte Chemie International Edition in English* 52, no. 22 (May).

Strecker, A., 1850, "Ueber die künstliche Bildung der Milchsäure und einen neuen, dem Glycocoll homologen Körper," *Annalen der Chemie und Pharmacie* 75, no. 1.

## Chapter 5. The Origin of Life: The Road to the Protocell

*Annotated Bibliography*

A seminal paper on the essence of cellular life and its potential origin on Earth is Szostak, Bartel, and Luisi (2001). The assembly, growth, and division of vesicles was experimentally studied, for example, by Hanczyc et al. (2003), Chen et al. (2004), Budin and Szostak (2011), Budin et al. (2014), and Kindt et al. (2020). The copying of RNA templates inside model protocells was examined by Adamala and Szostak (2013), and O'Flaherty et al. (2018). The Virtual Circular Genome model was tested by Ding et al. (2023). Various aspects of nonenzymatic assembly and replication were studied by Rajamani et al. (2010) and by Radakovic et al. (2022). The current state of the art in the laboratory evolution of RNA enzymes with RNA polymerase activity is beautifully illustrated in Papastavrou et al. (2024).

*References*

Adamala, K., and Szostak, J. W., 2013, "Nonenzymatic Template-Directed RNA Synthesis Inside Model Protocells," *Science* 342 (November 29).

Budin, I., and Szostak, J. W., 2011, "Physical Effects Underlying the Transition from Primitive to Modern Cell Membranes," *Proceedings of the National Academy of Sciences USA* 108, no. 14 (March).

Budin, I., Prywes, N., Zhang, N., and Szostak, J. W., 2014, "Chain-Length Heterogeneity Allows for the Assembly of Fatty Acid Vesicles in Dilute Solutions," *Biophysical Journal* 107, no. 7 (October).

Chen, I. A., Roberts, R. W., and Szostak, J. W., 2004, "The Emergence of Competition Between Model Protocells," *Science* 305 (September 3).

Ding, D., Zhou, L., Mittal, S., and Szostak, J. W., 2023, "Experimental Tests of the Virtual Circular Genome Model for Nonenzymatic RNA Replication," *Journal of the American Chemical Society* 145, no. 13 (April).

Hanczyc, M. M., Fujikawa, S. M., and Szostak, J. W., 2003, "Experimental Models of Primitive Cellular Compartments: Encapsulation, Growth, and Division," *Science* 302 (October 24).

Kindt, J., Szostak, J. W., and Wang, A., 2020, "Bulk Self-Assembly of Giant, Unilamellar Vesicles," *ACS Nano* 14, no. 11 (November).

O'Flaherty, D., Kamat, N. P., Mizra, F. N., Li, L., Prywes, N., and Szostak, J. W., 2018, "Copying of Mixed Sequence RNA Templates Inside Model Protocells," *Journal of the American Chemical Society* 140, no. 15 (April).

Papastavrou, N., Horning, D. P., and Joyce, G. F., 2024, "RNA-Catalyzed Evolution of Catalytic RNA," *Proceedings of the National Academy of Sciences USA* 121, no. 11 (March).

Radakovic, A., et al., 2022, "Nonenzymatic Assembly of Active Chimeric Ribozymes from Aminoacylated RNA Oligonucleotides," *Proceedings of the National Academy of Sciences USA* 119, no. 7 (February).

Rajamani, S., Ichida, J. K., Antal, T., Treco, D. A., Leu, K., Nowak, M. A., Szostak, J. W., and Chen, I. A., 2010, "Effect of Stalling After Mismatches on the Error Catastrophe in Nonenzymatic Nucleic Acid Replication," *Journal of the American Chemical Society* 132, no. 16 (April).

Szostak, J. W., Bartel, D. P., and Luigi Luisi, P., 2001, "Synthesizing Life," *Nature* 409 (January 18).

## Chapter 6. Putting It All Together: From Astrophysics and Geology to Chemistry and Biology

*Annotated Bibliography*

Sasselov, Grotzinger, and Sutherland (2020) showed how an integrative approach that combines laboratory experiments with geologic, geochemical, and astrophysical observations assists in the construction of a robust chemical pathway to life. Mann (2018) gave a brief popular description of the changing views on the reality (or not) of the Late Heavy Bombardment on Earth. A potential solution to the "faint young Sun paradox" and its implications were discussed in O'Callaghan (2022). Evidence further strengthening the idea that life on Earth may have arisen in shallow ponds of water on land, rather than at the bottom of oceans, was presented by Ranjan et al. (2019). In an interview published by *Quanta Magazine*, biochemist David Deamer also explained why he favors ponds as the place where life on Earth originated [see Singer (2016)]. The opposite opinion, suggesting that life originated at deep-sea hydrothermal vents, was reviewed and discussed by Russell (2021). Additional support for the shallow ponds scenario comes from the fact that Zhang et al. (2022) showed that simple and common environmental fluctuations of freeze-thaw cycles could have played an important role in prebiotic nucleotide activation, nonenzymatic RNA copying, and thereby the emergence of a genetic information system on the early Earth. The potential role of asteroid impacts on the early Earth in generating a reduced atmosphere, in bringing surface water, and in creating cradles for the origin of life was analyzed and discussed in Zahnle et al. (2020), in Osinski et al. (2020), and in Martin and Livio (2021, 2022).

*References*

Mann, A., 2018, "Bashing Holes in the Tale of Earth's Troubled Youth," *Nature* 553 (January 24).

Martin, R. G., and Livio, M., 2021, "How Much Water Was Delivered from the Asteroid Belt to the Earth After Its Formation?," *Monthly Notices of the Royal Astronomical Society* 506, no. 1 (September).

Martin, R. G., and Livio, M., 2022, "Asteroids and Life: How Special Is the Solar System?," *Astrophysical Journal Letters* 926, no. 2 (February).

O'Callaghan, J., 2022, "A Solution to the Faint Sun Paradox Reveals a Narrow Window for Life," *Quanta Magazine*, January 27, https://www.quantamagazine.org/the-sun-was-dimmer-when-earth-formed-how-did-life-emerge-20220127/.

Osinski, G. R., Cockell, C. S., Pontefract, A., and Sapers, H. M., 2020, "The Role of Meteorite Impacts in the Origin of Life," *Astrobiology* 20, no. 9 (September).

Ranjan, S., et al., 2019, "Nitrogen Oxide Concentrations in Natural Waters on Early Earth," *Geochemistry, Geophysics, Geosystems* 20, no. 4 (April).

Russell, M. J., 2021, "The 'Water Problem' (*sic*), Illusory Pond and Life's Submarine Emergence—A Review," *Life* 11, no. 5 (May).

Sasselov, D. D., Grotzinger, J. P., and Sutherland, J. D., 2020, "The Origin of Life as a Planetary Phenomenon," *Science Advances* 6, no. 6 (February).

Singer, E., 2016, "In Warm Greasy Puddles, the Spark of Life?," *Quanta Magazine*, March 17, https://www. quantamagazine.org/in-warm-greasy-puddles-the-spark-of-life-20160317/.

Zahnle, K. J., Lupu, R., Catling, D. C., and Wogan, N., 2020, "Creation and Evolution of Impact-Generated Reduced Atmospheres of Early Earth," *Planetary Science Journal* 1, no. 1 (May).

Zhang, S. J., Duzdevich, D., Ding, D., and Szostak, J. W., 2022, "Freeze-Thaw Cycles Enable a Prebiotically Possible and Continuous Pathway from Nucleotide Activation to Nonenzymatic RNA Copying," *Proceedings of the National Academy of Sciences USA* 119, no. 17 (April).

## Chapter 7. Extraterrestrial Life on Solar System Planets?

*Annotated Bibliography*

The search for life on Mars is beautifully and comprehensively described in Stewart Johnson (2020). Shindell (2023) presents the history of the human fascination with Mars. A richly illustrated and well explained presentation of the *Curiosity* mission to Mars is Kaufman (2014). The early stages in the story of the ALH84001 meteorite were presented by Sawyer (2006). The latest findings on the meteorite, which point to the organic molecules in the meteorite having been formed by geological (rather than biotic) processes, are presented in Steele et al. (2022). An earlier review of the meteorite data was published by Martel et al. (2012). The Viking experiments and their results are described in detail in Cooper (1980). The story of the Mars methane mystery is described in Shekhtman (2021). The question of whether plate tectonics are absolutely needed to sustain life was investigated by Foley and

Smye (2018). New evidence for volcanic activity on Mars was described and discussed by Plait (2023), and some of the results from the *Perseverance* rover were summarized by Chang (2022). The possibility of hot springs in the Gusev crater on Mars was presented by Ruff et al. (2020), and van Kranendonk et al. (2021) analyzed the phenomenon of hydrothermal fields in the context of the search for life in the solar system. A popular account of the tentative discovery of phosphine in the thick atmosphere of Venus and of its potential implications can be found in Stirone, Chang, and Overbye (2021). The scientific discovery itself was announced in Greaves et al. (2021). More technical discussions analyzing the result, and evaluating potential abiotic sources for the phosphine are Bains et al. (2021) and Bains et al. (2022).

*References*

Bains, W., et al., 2021, "Venusian Phosphine: A 'Wow!' Signal in Chemistry?," preprint, November 9, arXiv:2111.05182.

Bains, W., et al., 2022, "Constraints on the Production of Phosphine by Venusian Volcanoes," *Universe* 8, no. 1 (January).

Chang, K., 2022, "On Mars, a Year of Surprise and Discovery," *New York Times*, February 15, https://www.nytimes.com/2022/02/15/science/mars-nasa-perseverance.html.

Cooper Jr., H. S. F., 1980, *The Search for Life on Mars: Evolution of an Idea* (New York: Henry Holt and Company).

Foley, B. J., and Smye, A. J., 2018, "Carbon Cycling and Habitability of Earth-Sized Stagnant Lid Planets," *Astrobiology* 18, no. 7 (July).

Greaves, J. S., et al., 2021, "Phosphine Gas in the Cloud Decks of Venus," *Nature Astronomy* 5 (July).

Kaufman, M., 2014, *Mars Up Close: Inside the Curiosity Mission* (Washington, DC: National Geographic).

Martel, J., et al., 2012, "Biomimetic Properties of Minerals and the Search for Life in the Martian Meteorite ALH84001," *Annual Review of Earth and Planetary Sciences* 40.

Plait, P., 2023, "Volcanic Activity on Mars Upends Red Planet Assumptions," *Scientific American*, January 5, https://www.scientificamerican.com/article/volcanic-activity-on-mars-upends-red-planet-assumptions/.

Ruff, S. W., et al., 2020, "The Case for Ancient Hot Springs in Gusev Crater, Mars," *Astrobiology* 20, no. 4 (April).

Sawyer, K., 2006, *The Rock from Mars: A Detective Story on Two Planets* (New York: Random House).

Shekhtman, L., 2021, "First You See It, Then You Don't: Scientists Closer to Explaining Mars Methane Mystery," NASA Jet Propulsion Laboratory, June 29, https://www.jpl.nasa.gov/news/first-you-see-it-then-you-dont-scientists-closer-to-explaining-mars-methane-mystery.

Shindell, M., 2023, *For the Love of Mars: A Human History of the Red Planet* (Chicago: University of Chicago Press).

Steele, A., et al., 2022, "Organic Synthesis Associated with Serpentinization and Carbonation on Early Mars," *Science* 375 (January 13).

Stewart Johnson, S., 2020, *The Sirens of Mars: Searching for Life on Another World* (New York: Crown).

Stirone, S., Chang, K., and Overbye, D., 2021, "Life on Venus? Astronomers See a Signal in Its Clouds," *New York Times*, June 22, https://www.nytimes.com/2020/09/14/science/venus-life-clouds.html.

van Kranendonk, M. J., et al., 2021, "Terrestrial Hydrothermal Fields and the Search for Life in the Solar System," *Bulletin of the American Astronomical Society* 53, no. 4 (May).

**Chapter 8. Extraterrestrial Life on Solar System Moons?**

*Annotated Bibliography*

Hand (2020) gives an excellent description of the exploration of the moons of Jupiter and Saturn, and the search for life in the subsurface oceans of these moons. The potential habitability of Europa was originally discussed by Reynolds et al. (1983). The idea that life can be sustained in liquid oceans underneath very thick layers of ice is thought to receive support from studies of life-forms found in Lake Vostok in Antarctica [see Gura and Rogers (2020)]. Similar studies were carried out by John Priscu in Antarctica's Lake Mercer and Lake Whillans [see Nadis (2020)]. A good explanation at a popular level for why Enceladus has become one of the most attractive targets for the search for life in the solar system is provided by NASA (2017). The possibility of life on Titan was comprehensively discussed in McKay (2016), and Titan was described in detail in Lorenz (2020). The fact that the observed escape rates of methane from Enceladus were at least tentatively consistent with the hypothesis of habitable conditions for methanogens was described in Affholder et al. (2021). The discovery of hydrogen cyanide in a plume from Enceladus was described in Peter, Nordheim, and Hand (2023). Speculative ideas about the possibility of different types of life on Titan have been examined by Sandström and Rahm (2020), and the idea that small amounts of hydrogen cyanide may act to solvate polar molecules (water ice in particular) in liquid hydrocarbons was suggested by Lorenz, Lunine, and Neish (2011).

*References*

Affholder, A., et al., 2021, "Bayesian Analysis of Enceladus's Plume Data to Assess Methanogenesis," *Nature Astronomy* 5 (June).

Gura, C., and Rogers, S. O., 2020, "Metatranscriptomic and Metagenomic Analysis of Biological Diversity in Subglacial Lake Vostok (Antarctica)," *Biology* 9, no. 3 (March).

Hand, K. P., 2020, *Alien Oceans: The Search for Life in the Depths of Space* (Princeton: Princeton University Press).

Lorenz, R., 2020, *Saturn's Moon Titan: From 4.5 Billion Years Ago to the Present* (Sparkford, UK: Haynes Publishing).

Lorenz, R. D., Lunine, J. I., and Neish, C. D., 2011, "Cyanide Soap? Dissolved Material in Titan's Seas," European Planetary Science Congress (EPSC)—Division for Planetary Sciences meeting 2011, 488.

McKay, C. P., 2016, "Titan as the Abode of Life," *Life* 6, no. 1 (February).

Nadis, S., 2020, "He Found 'Islands of Fertility' Beneath Antarctica's Ice," *Quanta Magazine*, July 20, https://www.quantamagazine.org/john-priscu-finds-life-in-antarcticas-frozen-lakes-20200720/.

NASA, 2017, "The Moon with the Plume," April 12, http://solarsystem.nasa.gov/news/13020/the-moon-with-the-plume/.

Peter, J. S., Nordheim, T. A., and Hand, K. P., 2023, "Detection of HCN and Diverse Redox Chemistry in the Plume of Enceladus," *Nature Astronomy* 8.

Reynolds, R. T., Squires, S. W., Colburn, D. S., and McKay, C. P., 1983, "On the Habitability of Europa," *Icarus* 56, no. 2 (November).

Sandström, H., and Rahm, M., 2020, "Can Polarity-Inverted Membranes Self-Assemble on Titan?," *Science Advances* 6, no. 4 (January).

**Chapter 9. Life Out There: The Astronomical Quest**

*Annotated Bibliography*

The literature on detecting exoplanets in general, and on habitable planets and their properties in particular, is vast. A beautiful description of the search for Earth-like planets is Kaltenegger (2024). A summary of some of the methods involved and characteristics of exoplanets is Mason (2010). Perhaps the most comprehensive reference for exoplanet research is Perryman (2018). Another book, emphasizing the great diversity of exoplanets, is Summers and Trefil (2017). The number of habitable exoplanets is estimated in Dressing and Charbonneau (2015), and in Bryson et al. (2021). Schulze-Makuch et al. (2020) identify a few exoplanets that in their opinion are even better for life than Earth. Hill et al. (2023) present a catalog of exoplanets in the habitable zone of their host stars. The topic of niche construction is explained in Laland, Mathews, and Feldman (2016). An interesting interview with computer modeler of potentially habitable worlds Lisa Kaltenegger can be found in Sokol (2022). Exoplanet atmospheres and methods for studying them are described in Seager and Deming (2010) and in Deming and Seager (2017). Excellent, detailed (technical) discussions of potential exoplanet biosignatures are presented, for example, in Catling et al. (2018) and in Schwieterman et al. (2018). The more general topic of what constitutes a biosignature is reviewed in Chan et al. (2019). The specific topic of oxygen as a biosignature is carefully examined by Meadows et al. (2018). The feasibility of detecting ocean glint on exoplanets was analyzed by Lustig-Yaeger et al. (2018). The habitability of planets around M-dwarf stars has been extensively debated. Examples of the discussions concerning various aspects of the problems involved in planets orbiting low-mass stars can be found in Shields, Ballard, and Asher Johnson (2016), Wandel (2018), Ranjan, Wordsworth, and Sasselov (2017), and Childs, Martin, and Livio (2022). A list of potentially habitable exoplanets is maintained and updated at the PHL@UPR Arecibo "Habitable Exoplanets Catalog": https://phl.upr.edu/projects/habitable-exoplanets-catalog. The possibility of having life on exomoons (moons orbiting exoplanets) is nicely discussed in a popular article by Billings (2017). A few of the recent observations of TRAPPIST-1 exoplanets are described in Greene et al. (2023) and Zieba at al. (2023), and their potential implications are discussed in Selsis et al. (2023). The spectra of flares from TRAPPIST-1 were studied by Howard et al. (2023). The detection of methane and carbon dioxide, and the suggested potential detection of dimethyl sulfide in the atmosphere of K2-18b, is described in Madhusudhan et al. (2023). Shorttle et al. (2024) suggest that K2-18b may have a molten surface rather than a water ocean. Recommendations from the 2020 Astrophysics Decadal Survey can be found in Dreier (2021) and Kaufman (2021).

*References*

Billings, L., 2017, "The Search for Life on Faraway Moons," *Scientific American*, special ed. (Fall 2017).

Bryson, S., et al., 2021, "The Occurrence of Rocky Habitable-Zone Planets Around Solar-Like Stars from Kepler Data," *Astronomical Journal* 161, no. 1 (December).

Catling, D. C., et al., 2018, "Exoplanet Biosignatures: A Framework for Their Assessment," *Astrobiology* 18, no. 6 (June).

Chan, M. A., et al., 2019, "Deciphering Biosignatures in Planetary Contexts," *Astrobiology* 19, no. 9 (September).

Childs, A. C., Martin, R. G., and Livio, M., 2022, "Life on Exoplanets in the Habitable Zone of M Dwarfs?," *Astrophysical Journal Letters* 937, no. 2 (October).

Deming, L. D., and Seager, S., 2017, "Illusion and Reality in the Atmospheres of Exoplanets," *Journal of Geophysical Research: Planets* 122.

Dreier, C., 2021, "Your Guide to the 2020 Astrophysics Decadal Survey: The Future, If You Want It," Planetary Society, December 3, https://www.planetary.org/articles/the-2020-astrophysics-decadal-survey-guide.

Dressing, C. D., and Charbonneau, D., 2015, "The Occurrence of Potentially Habitable Planets Orbiting M Dwarfs Estimated from the Full Kepler Dataset and an Empirical Measurement of the Detection Sensitivity," *Astrophysical Journal* 807, no. 1 (July).

Greene, T. P., et al., 2023, "Thermal Emission from the Earth-Sized Exoplanet TRAPPIST-1 b Using JWST," *Nature* 618 (June 1).

Hill, M. L., et al., 2023, "A Catalog of Habitable Zone Exoplanets," *Astronomical Journal* 165, no. 34 (February).

Howard, W. S., et al., 2023, "Characterizing the Near-Infrared Spectra of Flares from TRAPPIST-1 During JWST Transit Spectroscopy Observations," October 5, arXiv:2310.03792.

Kaltenegger, L., 2024, *Alien Earths: The New Science of Planet Hunting in the Cosmos* (New York: St. Martin's Press).

Kaufman, M., 2021, "NASA Should Build a Grand Observatory Designed to Search for Life Beyond Earth, Panel Concludes," Many Worlds, November 5, https://manyworlds.space/2021/11/05/nasa

-should-build-a-grand-observatory-designed-to-search-for-life-beyond-earth-panel-concludes/.

Laland, K., Matthews, B., and Feldman, M. W., 2016, "An Introduction to Niche Construction Theory," *Evolutionary Ecology* 30, no. 2.

Lustig-Yaeger, J., et al., 2018, "Detecting Ocean Glint on Exoplanets Using Multiphase Mapping," *Astronomical Journal* 156, no. 6 (December).

Madhusudhan, N., et al., 2023, "Carbon-Bearing Molecules in a Possible Hycean Atmosphere," October 4, arXiv:2309.05566.

Mason, J. W., editor, 2010, *Exoplanets: Detection, Formation, Properties, Habitability* (Chichester, UK: Praxis Publishing).

Meadows, V. S., et al., 2018, "Exoplanet Biosignatures: Understanding Oxygen as a Biosignature in the Context of Its Environment," *Astrobiology* 18, no. 6 (June).

Perryman, M., 2018, *The Exoplanet Handbook*, 2nd ed. (Cambridge: Cambridge University Press).

Ranjan, S., Wordsworth, R., and Sasselov, D., 2017, "The Surface UV Environment on Planets Orbiting M Dwarfs: Implications for Prebiotic Chemistry and the Need for Experimental Follow-Up," *Astrophysical Journal* 843, no. 110 (July).

Schulze-Makuch, D., Heller, R., and Guinan, E., 2020, "In Search for a Planet Better Than Earth: Top Contenders for a Superhabitable World," *Astrobiology* 20, no. 12 (December).

Schwieterman, E. W., et al., 2018, "Exoplanet Biosignatures: A Review of Remotely Detectable Signs of Life," *Astrobiology* 18, no. 6 (June).

Seager, S., and Deming, D., 2010, "Exoplanet Atmospheres," *Annual Review of Astronomy and Astrophysics* 48.

Selsis, F., et al., 2023, "A Cool Runaway Greenhouse Without Surface Magma Ocean," *Nature* 620 (August 9).

Shields, A. L., Ballard, S., and Asher Johnson, J., 2016, "The Habitability of Planets Orbiting M-Dwarf Stars," *Physics Reports* 663 (December).

Shorttle, O., et al., 2024, "Distinguishing Oceans of Water from Magma on Mini-Neptune K2-18b," *Astrophysical Journal Letters* 962, no. 1 (February).

Sokol, J., 2022, "A Dream of Discovering Alien Life Finds New Hope," *Quanta Magazine,* November 3, https://www.quantamagazine.org/alien-life-a-dream-of-discovery-finds-new-hope-20221103/.

Summers, M., and Trefil, J., 2017, *Exoplanets: Diamond Worlds, Super Earths, Pulsar Planets, and the New Search for Life Beyond Our Solar System* (Washington, DC: Smithsonian Books).

Wandel, A., 2018, "On the Biohabitability of M-Dwarf Planets," *Astronomical Journal* 856, no. 3 (April).

Zieba, S., et al., 2023, "No Thick Carbon Dioxide Atmosphere on the Rocky Exoplanet TRAPPIST-1 c," *Nature* 620 (August 24).

## Chapter 10. Life as We Don't Know It: The Design of Natural and Unnatural Life-Forms

*Annotated Bibliography*

Szostak (2017) discusses the fact that the broad scope of synthetic chemistry suggests it may be possible to design artificial life based on a biochemistry that is different from that of life on Earth. Moskowitz (2017) gives a nice popular account of extreme molecules being detected in space. An excellent study of silicon as a potential building block for life (instead of carbon) is Petkowski, Bains, and Seager (2020). Bains et al. (2021) examined the possibility of using sulfuric acid as an alternative solvent to water. Kunieda, Nakamura, and Evans (1991) studied the production of inside-out membrane vesicles in nonpolar organic solvents. The formation of artificial (chemically different from biological) membranes was studied by Hardy et al. (2015). Scoles (2023) gives a popular account of the search for life as we don't know it. Alternative genetic codes are briefly reviewed by Kubyshkin and Budisa (2019). In an imaginative, speculative book, Trefil and Summers (2019) explore possible answers to the question of what alien life might look like. An interesting attempt to identify universal scaling laws across biochemical reactions used by life is Gagler et al. (2022). Bostrom (2014) gives an excellent discussion of the topic of a potential AI-based life and "superintelligence."

*References*

Bains, W., Petkowski, J. J., Zhan, Z., and Seager, S., 2021, "Evaluating Alternatives to Water as Solvents for Life: The Example of Sulfuric Acid," *Life* 11, no. 5 (May).

Bostrom, N., 2014, *Superintelligence: Paths, Dangers, Strategies* (Oxford: Oxford University Press).

Hardy, M. D., et al., 2015, "Self-Reproducing Catalyst Drives Repeated Phospholipid Synthesis and Membrane Growth," *Proceedings of the National Academy of Sciences USA* 112, no. 27 (July).

Gagler, D. C., Karas, B., Kempes, C. P., and Waker, S. I., 2022, "Scaling Laws in Enzyme Function Reveal a New Kind of Biochemical Universality," *Proceedings of the National Academy of Sciences USA* 119, no. 9 (March).

Kubyshkin, V., and Budisa, N., 2019, "Anticipating Alien Cells with Alternative Genetic Codes: Away from the Alanine World!," *Current Opinion in Biotechnology* 60 (December).

Kunieda, H., Nakamura, K., and Evans, D. F., 1991, "Formation of Reversed Vesicles," *Journal of the American Chemical Society* 113, no. 3.

Moskowitz, C., 2017, "Extreme Molecules in Space," *Scientific American*, special ed. (Fall 2017).

Petkowski, J. J., Bains, W., and Seager, S., 2020, "On the Potential of Silicon as a Building Block for Life," *Life* 10, no. 6 (June).

Scoles, S., 2023, "The Search for Extraterrestrial Life as We Don't Know It," *Scientific American*, February 1.

Szostak, J. W., 2017, "The Origin of Life on Earth and the Design of Alternative Life Forms," *Molecular Frontiers Journal* 1, no. 2 (December).

Trefil, J., and Summers, M., 2019, *Imagined Life: A Speculative Scientific Journey Among the Exoplanets in Search of Intelligent Aliens, Ice Creatures, and Supergravity Animals* (Washington, DC: Smithsonian Books).

## Chapter 11. The Hunt for Intelligence: Preliminary Thoughts

*Annotated Bibliography*

The Drake equation has been extensively discussed. Examples are in Lemonick (1998) and Frank (2018). The suggestion that alien intelligent civilizations probably existed somewhere in the universe is also examined in Frank (2018). The different version of the Drake equation, which gives an estimate for the number of exoplanets with detectable biosignatures from a combination of observations by TESS and JWST, is discussed in Seager (2016). Seventy-five possible solutions to the Fermi Paradox are presented, analyzed, and discussed in Webb (2015). A popular, brief review of the search for extraterrestrial life is Livio and Silk (2017). An empirical approach attempting to estimate the number of technological species in the Milky Way is presented by Engler and von Wehrden (2019). Arguments that complex life is rare, and that we may be alone in the cosmos, are presented in Carter (1983), Ward and Brownlee (2000), Davies (2010), and Gribbin (2011). Livio (1999) presented a potential weakness in Carter's argument. The Anthropic Principle is comprehensibly discussed by Bostrom (2002). Livio (2023) gives a popular-level discussion of some implications of the Copernican Principle.

*References*

Bostrom, N., 2002, *Anthropic Bias: Observation Selection Effects in Science and Philosophy* (Abingdon: Routledge).

Carter, B., 1983, "The Anthropic Principle and Its Implications for Biological Evolution," *Philosophical Transactions of the Royal Society of London A* 310 (December).

Davies, P., 2010, *The Eerie Silence: Renewing Our Search for Alien Intelligence* (Boston: Houghton Mifflin Harcourt).

Engler, J.-O., and von Wehrden, H., 2019, "'Where Is Everybody?': An Empirical Appraisal of Occurrence, Prevalence and Sustainability of Technological Species in the Universe," *International Journal of Astrobiology* 18, no. 6.

Frank, A., 2018, *Light of the Stars: Alien Worlds and the Fate of the Earth* (New York: W. W. Norton).

Gribbin, J., 2011, *Alone in the Universe: Why Our Planet Is Unique* (Hoboken, NJ: John Wiley & Sons).

Lemonick, M. D., 1998, *Other Worlds: The Search for Life in the Universe* (New York: Simon & Schuster).

Livio, M., 1999, "How Rare Are Extraterrestrial Civilizations, and When Did They Emerge?," *Astrophysical Journal* 511, no. 1 (January).

Livio, M., 2023, "How Far Should We Take Our Cosmic Humility?," *Scientific American*, April 19, https://www.scientificamerican.com/article/how-far-should-we-take-our-cosmic-humility1/.

Livio, M., and Silk, J., 2017, "Where Are They?," *Physics Today* 70, no. 3 (March).

Seager, S., 2016, "Are They Out There? Technology, the Drake Equation, and Looking for Life on Other Worlds," in Al-Khalili, J., editor, 2016, *Aliens: The World's Leading Scientists on the Search for Extraterrestrial Life* (New York: Picador), 188.

Ward, P. D., and Brownlee, D., 2000, *Rare Earth: Why Complex Life Is Uncommon in the Universe* (New York: Copernicus).

Webb, S., 2015, *If the Universe Is Teeming with Aliens . . . WHERE IS EVERYBODY? Seventy-Five Solutions to the Fermi Paradox and the Problem of Extraterrestrial Life* (New York: Springer).

## Chapter 12. The Hunt for Intelligence: The Searches

*Annotated Bibliography*

A nice description of the early SETI searches is Shostak (2009). A recent, delightful book on the search for aliens is Frank (2023). An examination of SETI science that also follows one of its pioneers, Jill

Tarter, is Scoles (2017). The SETI searches as a "needle in a haystack" metaphor are discussed, for example, in Tarter et al. (2010). A more recent evaluation of the SETI efforts is Wright, Kanodia, and Lubar (2018). Loeb's controversial suggestion that the interstellar object Oumuamua was a piece of advanced technology created by an alien civilization is extensively described in Loeb (2021). The results from the analysis of spherules from a meteorite is in Loeb et al. (2023). Siegel (2023) clearly presents the case that the spherules' composition may be simply industrial pollutants. This is disputed in Loeb et al. (2024). The strange behavior of Boyajian's Star and possible explanations were discussed, for example, by Cartier and Wright (2017). The conclusive analysis of the signal dubbed BLC1 was published in Sheikh et al. (2021). Rees and Livio (2023) discuss the possibility that most technological species in our galaxy may be AI machines.

*References*

Cartier, K., and Wright, J. T., 2017, "Strange News from Another Star," *Scientific American*, special ed. (Fall 2017).

Frank, A., 2023, *The Little Book of Aliens* (New York: HarperCollins).

Loeb, A., 2021, *Extraterrestrial: The First Sign of Intelligent Life Beyond Earth* (Boston: Houghton Mifflin Harcourt).

Loeb, A., et al., 2023, "Discovery of Spherules of Likely Extrasolar Composition in the Pacific Ocean Site of the CNEOS 2014-01-08 (IM1) Bolide," preprint, August 9, arXiv:2308.15623.

Loeb, A., et al., 2024, "Recovery and Classification of Spherules from the Pacific Ocean Site of the CNEOS 2014 January 8 (IM1) Bolide," *Research Notes of the AAS* 8, no. 1 (January).

Rees, M., and Livio, M., 2023, "Most Aliens May Be Artificial Intelligence, Not Life as We Know It," *Scientific American*, June 1, https://www.scientificamerican.com/article/most-aliens-may-be-artificial-intelligence-not-life-as-we-know-it/.

Scoles, S., 2017, *Making Contact: Jill Tarter and the Search for Extraterrestrial Intelligence* (New York: Pegasus Books).

Sheikh, S. Z., et al., 2021, "Analysis of the Breakthrough Listen Signal of Interest blc1 with a Technosignature Verification Framework," *Nature Astronomy* 5.

Shostak, S., 2009, *Confessions of an Alien Hunter: A Scientist's Search for Extraterrestrial Intelligence* (Washington, DC: National Geographic).

Siegel, E., 2023, "Harvard Astronomer's 'Alien Spherules' Are Industrial Pollutants," Big Think, November 14, https://bigthink.com/starts-with-a-bang/harvard-astronomer-alien-spherules/.

Tarter, J. C., et al., 2010, "SETI Turns 50: Five Decades of Progress in the Search for Extraterrestrial Intelligence," *Proceedings of the SPIE* 7819 (August).

Wright, J. T., Kanodia, S., and Lubar, E., 2018, "How Much SETI Has Been Done? Finding Needles in the *n*-Dimensional Cosmic Haystack," *Astronomical Journal* 156, no. 6 (November).

## Chapter 13. Epilogue: An Imminent Breakthrough?

*Annotated Bibliography*

To get an idea of how much our knowledge about the universe has progressed during the past century, Arrhenius (1909) provides for a fascinating read. The science that could be achieved with a new UV spectrograph is described on the web page of the Habitable Worlds Observatory in a white paper by Tumlinson et al.

*References*

Arrhenius, S., 1909, *The Life of the Universe: As Conceived by Man from the Earliest Ages to the Present Time* (London: Harper & Brothers).

Tumlinson, J., et al., "Unique Astrophysics in the Lyman Ultraviolet," https://www.stsci.edu/~tumlinso/LymanUV-Tumlinson.pdf.

# INDEX

Credit: Johnny Boston

**Mario Livio** is an astrophysicist who worked with the Hubble Space Telescope. He is a best-selling author of seven books, including *The Golden Ratio* and *Brilliant Blunders*. He lives in Hoboken, New Jersey.

Credit: Photo courtesy of the author

**Jack Szostak** is a professor of chemistry at the University of Chicago, leading the Center for the Origin of Life. He shared the 2009 Nobel Prize for Physiology or Medicine. He lives in Chicago, Illinois.